Polynomials

Polynomials

Theory and Applications

Editor

Cheon Seoung Ryoo

MDPI • Basel • Beijing • Wuhan • Barcelona • Belgrade • Manchester • Tokyo • Cluj • Tianjin

Editor
Cheon Seoung Ryoo
Department of Mathematics,
Hannam University
Korea

Editorial Office
MDPI
St. Alban-Anlage 66
4052 Basel, Switzerland

This is a reprint of articles from the Special Issue published online in the open access journal *Mathematics* (ISSN 2227-7390) (available at: https://www.mdpi.com/journal/mathematics/special_issues/Polynomials_Theory_Applications).

For citation purposes, cite each article independently as indicated on the article page online and as indicated below:

LastName, A.A.; LastName, B.B.; LastName, C.C. Article Title. *Journal Name* **Year**, *Article Number*, Page Range.

ISBN 978-3-03943-314-8 (Hbk)
ISBN 978-3-03943-315-5 (PDF)

Contents

About the Editor

Cheon Seoung Ryoo is a professor in Mathematics at Hannam University. He received his Ph.D. degree in Mathematics from Kyushu University. Dr. Ryoo is the author of several research articles on numerical computations with guaranteed accuracy. He has also made contributions to the field of scientific computing, p-adic functional analysis, and analytic number theory. More recently, he has been working with differential equations, dynamical systems, quantum calculus, and special functions.

Preface to "Polynomials"

The importance of polynomials in the interdisciplinary field of mathematics, engineering, and science is well known. Over the past several decades, research on polynomials has been conducted extensively in many disciplines.

This book is a collection of selected and refereed papers on the most recent results concerning polynomials in mathematics, science, and industry. There are so many topics related to polynomials that it is hard to list them all; some relevant topics include but are not limited to the following:

- The modern umbral calculus
- Orthogonal polynomials, matrix orthogonal polynomials, multiple orthogonal polynomials
- Matrix and determinant approach to special polynomial sequences
- Applications of special polynomial sequences
- Number theory and special functions
- Asymptotic methods in orthogonal polynomials
- Fractional calculus and special functions
- Symbolic computations and special functions
- Applications of special functions to statistics, physical sciences, and engineering

We hope that this book is timely and will fill a gap in the literature on the theory of polynomials and related fields. We also hope that it will promote further research and development in this important area.

We also thank the authors for their creative contributions and the reviewers for their prompt and careful reviews.

Cheon Seoung Ryoo
Editor

 *mathematics*

Article

A Parametric Kind of the Degenerate Fubini Numbers and Polynomials

Sunil Kumar Sharma [1],*, Waseem A. Khan [2] and Cheon Seoung Ryoo [3]

[1] College of Computer and Information Sciences, Majmaah University, Majmaah 11952, Saudi Arabia
[2] Department of Mathematics and Natural Sciences, Prince Mohammad Bin Fahd University, P.O Box 1664, Al Khobar 31952, Saudi Arabia; wkhan1@pmu.edu.sa
[3] Department of Mathematics, Hannam University, Daejeon 34430, Korea; ryoocs@hnu.kr
* Correspondence: s.sharma@mu.edu.sa

Received: 26 February 2020; Accepted: 7 March 2020; Published: 12 March 2020

Abstract: In this article, we introduce the parametric kinds of degenerate type Fubini polynomials and numbers. We derive recurrence relations, identities and summation formulas of these polynomials with the aid of generating functions and trigonometric functions. Further, we show that the parametric kind of the degenerate type Fubini polynomials are represented in terms of the Stirling numbers.

Keywords: Fubini polynomials; degenerate Fubini polynomials; Stirling numbers

MSC: 11B83; 11C08; 11Y35

---

1. Introduction

In the last decade, many mathematicians, namely, Kargin [1], Duran and Acikgoz [2], Kim et al. [3,4], Kilar and Simsek [5], Su and He [6] have been studied in the area of the Fubini polynomials and numbers, degenerate Fubini polynomials and numbers. The range of Appell polynomials sequences is one of the important classes of polynomial sequences. The Appell polynomials are very frequently used in various problems in pure and applied mathematics related to functional equations in differential equations, approximation theories, interpolation problems, summation methods, quadrature rules, and their multidimensional extensions (see [7,8]). The sequence of Appell polynomials $A_j(w)$ can be signified by means either following equivalent conditions

$$\frac{d}{dw}A_j(w) = jA_{j-1}(w), \quad A_0(w) \neq 0, w = \eta + i\xi \in \mathbb{C}, \quad j \in \mathbb{N}_0, \tag{1}$$

and satisfying the generating function

$$A(z)e^{wz} = A_0(w) + A_1(w)\frac{z}{1!} + A_2(w)\frac{z^2}{2!} + \cdots + A_n(w)\frac{z^n}{n!} + \cdots = \sum_{j=0}^{\infty} A_j(w)\frac{z^j}{j!}, \tag{2}$$

where $A(w)$ is an entirely real power series with Taylor expansion given by

$$A(w) = A_0(w) + A_1(w)\frac{z}{1!} + A_2(w)\frac{z^2}{2!} + \cdots + A_j(w)\frac{z^j}{j!} + \cdots, \quad A_0 \neq 0.$$

The well known degenerate exponential function [9] is defined by

$$e_\mu^\eta(z) = (1 + \mu z)^{\frac{\eta}{\mu}}, \quad e_\mu(z) = e_\mu^1(z), (\mu \in \mathbb{R}). \tag{3}$$

Since

$$\lim_{\mu \to 0} (1 + \mu z)^{\frac{\eta}{\mu}} = e^{\eta z}.$$

In [10,11], Carlitz introduced the degenerate Bernoulli polynomials which are defined by

$$\frac{z}{(1 + \mu z)^{\frac{1}{\mu}} - 1} (1 + \mu z)^{\frac{\eta}{\mu}} = \sum_{j=0}^{\infty} \beta_j(\eta; \mu) \frac{z^j}{j!}, \, (\mu \in \mathbb{C}), \tag{4}$$

so that

$$\beta_j(\eta; \mu) = \sum_{w=0}^{j} \binom{j}{w} \beta_w(\mu) \left(\frac{\eta}{\mu} \right)_{j-w}. \tag{5}$$

When $\eta = 0$, $\beta_j(\mu) = \beta_j(0; \mu)$ are called the degenerate Bernoulli numbers, (see [12–15]).

From Equation (4), we get

$$\begin{aligned}
\sum_{j=0}^{\infty} \lim_{\mu \to 0} \beta_j(\eta; \mu) \frac{z^j}{j!} &= \lim_{\mu \to 0} \frac{z}{(1+\mu z)^{\frac{1}{\mu}} - 1} (1 + \mu z)^{\frac{\eta}{\mu}} \\
&= \frac{z}{e^z - 1} e^{\eta z} = \sum_{j=0}^{\infty} B_j(\eta) \frac{z^j}{j!},
\end{aligned} \tag{6}$$

where $B_j(\eta)$ are called the Bernoulli polynomials, (see [9,16]).

The Stirling numbers of the first kind [3,14,17]) are defined by

$$\eta^j = \sum_{i=0}^{j} S_1(j, i)(\eta)_i, \, (j \geq 0), \tag{7}$$

where $(\eta)_0 = 1, (\eta)_j = \eta(\eta - 1) \cdots (\eta - j + 1), (j \geq 1)$. Alternatively, the Stirling numbers of the second kind are defined by following generating function (see [4,5])

$$\frac{(e^z - 1)^j}{j!} = \sum_{i=j}^{\infty} S_2(j, i) \frac{z^j}{j!}. \tag{8}$$

The degenerate Stirling numbers of the second kind [17] are defined by means of the following generating function

$$\frac{1}{i!} \left((1 + \mu z)^{\frac{1}{\mu}} - 1 \right)^i = \sum_{j=i}^{\infty} S_{2,\mu}(j, i) \frac{z^j}{j!}. \tag{9}$$

It is clear that

$$\lim_{\mu \to 0} S_{2,\mu}(j, i) = S_2(j, i).$$

The generating function of 2-variable degenerate Fubini polynomials [3] are defined by

$$\frac{1}{1 - \xi((1 + \mu z)^{\frac{1}{\mu}} - 1)} (1 + \mu z)^{\frac{\eta}{\mu}} = \sum_{j=0}^{\infty} F_{j,\mu}(\eta; \xi) \frac{z^j}{j!}, \tag{10}$$

so that

$$F_{j,\mu}(\eta; \xi) = \sum_{i=0}^{j} \binom{j}{i} F_{i,\mu}(\xi)(\eta)_{j-i,\mu}.$$

When $\eta = 0$, $F_{j,\mu}(0; \xi) = F_{j,\mu}(\xi)$, $F_{j,\mu}(0; 1) = F_{j,\mu}$ are called the degenerate Fubini polynomials and degenerate Fubini numbers.

Note that

$$\lim_{\mu \to 0} \sum_{j=0}^{\infty} F_{j,\mu}(\eta;\xi)\frac{z^j}{j!} = \lim_{\mu \to 0} \frac{1}{1-\xi((1+\mu z)^{\frac{1}{\mu}}-1)}(1+\mu z)^{\frac{\eta}{\mu}}$$

$$= \frac{1}{1-\xi(e^z-1)}e^{\eta z} = \sum_{j=0}^{\infty} F_j(\eta;\xi)\frac{z^j}{j!}, \tag{11}$$

where $F_j(\eta;\xi)$ are called the 2-variable Fubini polynomials, (see, [1,18]).

The two trigonometric functions $e^{\eta z}\cos \xi z$ and $e^{\eta z}\sin \xi z$ are defined as follows (see [19,20]):

$$e^{\eta z}\cos \xi z = \sum_{k=0}^{\infty} C_k(\eta,\xi)\frac{z^k}{k!}, \tag{12}$$

and

$$e^{\eta z}\sin \xi z = \sum_{k=0}^{\infty} S_k(\eta,\xi)\frac{z^k}{k!}, \tag{13}$$

where

$$C_k(\eta,\xi) = \sum_{j=0}^{[\frac{k}{2}]} \binom{k}{2j}(-1)^j \eta^{k-2j}\xi^{2j}, \tag{14}$$

and

$$S_k(\eta,\xi) = \sum_{j=0}^{[\frac{k-1}{2}]} \binom{k}{2j+1}(-1)^j \eta^{k-2j-1}\xi^{2j+1}. \tag{15}$$

Recently, Kim et al. [16] introduced the degenerate cosine-polynomials and degenerate sine-polynomials are respectively, as follows

$$e_\mu^\eta(z)\cos_\lambda^\xi(z) = \sum_{j=0}^{\infty} C_{j,\mu}(\eta,\xi)\frac{z^j}{j!}, \tag{16}$$

and

$$e_\mu^\eta(z)\sin_\mu^\xi(z) = \sum_{j=0}^{\infty} S_{j,\mu}(\eta,\xi)\frac{z^j}{j!}, \tag{17}$$

where

$$C_{j,\mu}(\eta,\xi) = \sum_{k=0}^{[\frac{j}{2}]} \sum_{i=2k}^{j} \binom{j}{i}\mu^{i-2k}(-1)^k \xi^{2k} S^1(i,2k)(\eta)_{j-i,\mu}, \tag{18}$$

and

$$S_{j,\mu}(\eta,\xi) = \sum_{k=0}^{[\frac{i-1}{2}]} \sum_{i=2k+1}^{j} \binom{j}{i}\mu^{i-2k-1}(-1)^k \xi^{2k+1} S^1(i,2k+1)(\eta)_{j-i,\mu}. \tag{19}$$

This paper is organized as follows: In Section 2, we introduce degenerate complex Fubini polynomials with degenerate cosine-Fubini and degenerate sine-Fubini polynomials and present some properties and their relations. In Section 3, we derive partial differentiation, recurrence relations and summation formulas, Stirling numbers of the second kind of degenerate type Fubini numbers and polynomials by using a generating function, respectively.

2. A Parametric Kind of the Degenerate Fubini Polynomials

In this section, we study the parametric kind of degenerate Fubini polynomials by employing the real and imaginary parts separately and introduce the degenerate Fubini polynomials in terms of degenerate complex polynomials.

The well known degenerate Euler's formula is defined as follows (see [16])

$$e_\mu^{(\eta+i\xi)z} = e_\mu^{\eta z} e_\mu^{i\xi z} = e_\mu^{\eta z}(\cos_\mu \xi z + i \sin_\mu \xi z), \tag{20}$$

where

$$\cos_\mu z = \frac{e_\mu^i(z) + e_\mu^{-i}(z)}{2}, \quad \sin_\mu z = \frac{e_\mu^i(z) - e_\mu^{-i}(z)}{2i}. \tag{21}$$

Note that

$$\lim_{\mu\to 0} e_\mu^i = e^{iz}, \lim_{\mu\to 0} \cos_\mu z = \cos z, \quad \lim_{\mu\to 0} \sin_\mu z = \sin z.$$

Using (8), we find

$$\frac{1}{1-\rho(e_\mu(z)-1)} e_\mu^{\eta+i\xi}(z) = \sum_{j=0}^\infty F_{j,\mu}(\eta+i\xi;\rho)\frac{z^j}{j!}, \tag{22}$$

and

$$\frac{1}{1-\rho(e_\mu(z)-1)} e_\mu^{\eta-i\xi}(z) = \sum_{j=0}^\infty F_{j,\mu}(\eta+i\xi;\rho)\frac{z^j}{j!}. \tag{23}$$

From Equations (22) and (23), we obtain

$$\frac{1}{1-\rho(e_\mu(z)-1)} e_\mu(\eta z)\cos_\mu(\xi z) = \sum_{j=0}^\infty \frac{F_{j,\mu}(\eta+i\xi;\rho) + F_{j,\mu}(\eta-i\xi;\rho)}{2}\frac{z^j}{j!}, \tag{24}$$

and

$$\frac{1}{1-\rho(e_\mu(z)-1)} e_\mu(\eta z)\sin_\mu(\xi z) = \sum_{j=0}^\infty \frac{F_{j,\mu}(\eta+i\xi;\rho) - F_{j,\mu}(\eta-i\xi;\rho)}{2i}\frac{z^j}{j!}. \tag{25}$$

Definition 1. *For a non negative integer n, let us define the degenerate cosine-Fubini polynomials $F_{j,\mu}^{(c)}(\eta,\xi;\rho)$ and the degenerate sine-Fubini polynomials $F_{j,\mu}^{(s)}(\eta,\xi;\rho)$ by the generating functions, respectively, as follows*

$$\frac{1}{1-\rho(e_\mu(z)-1)} e_\mu(\eta z)\cos_\mu(\xi z) = \sum_{j=0}^\infty F_{j,\mu}^{(c)}(\eta,\xi;\rho)\frac{z^j}{j!}, \tag{26}$$

and

$$\frac{1}{1-\rho(e_\mu(z)-1)} e_\mu(\eta z)\sin_\mu(\xi z) = \sum_{j=0}^\infty F_{j,\mu}^{(s)}(\eta,\xi;\rho)\frac{z^j}{j!}. \tag{27}$$

It is noted that

$$F_{j,\mu}^{(c)}(0,0;1) = F_{j,\mu}, \quad F_{j,\mu}^{(s)}(0,0;1) = 0, (j \geq 0).$$

The first few of them are:

$$F_{0,\mu}^{(c)}(\eta,\xi;\rho) = 1,$$
$$F_{1,\mu}^{(c)}(\eta,\xi;\rho) = \eta + \rho,$$
$$F_{2,\mu}^{(c)}(\eta,\xi;\rho) = -\mu\eta + \eta^2 - \xi^2 + \rho - \mu\rho + 2\eta\rho + 2\rho^2,$$
$$F_{3,\mu}^{(c)}(\eta,\xi;\rho) = 2\mu^2\eta - 3\mu\eta^2 + \eta^3 - 3\eta\xi^2 + 3\mu\xi^3 + \rho - 3\mu\rho + 2\mu^2\rho + 3\eta\rho - 6\mu\eta\rho$$
$$+ 3\eta^2\rho - 3\xi^2\rho + 6\rho^2 - 6\mu\rho^2 + 6\eta\rho^2 + 6\rho^3,$$

and

$$F_{0,\mu}^{(s)}(\eta,\xi;\rho) = 0,$$

$$F_{1,\mu}^{(s)}(\eta,\xi;\rho) = \xi,$$

$$F_{2,\mu}^{(s)}(\eta,\xi;\rho) = 2\eta\xi - \mu\xi^2 + 2\xi\rho,$$

$$F_{3,\mu}^{(s)}(\eta,\xi;\rho) = -3\mu\eta\xi + 3\eta^2\xi - 3\mu\eta\xi^2 - \xi^3 + 2\mu^2\xi^3 + 3\xi\rho - 3\mu\xi\rho$$
$$+ 6\eta\xi\rho - 3\mu\xi^2\rho + 6\xi\rho^2.$$

Note that $\lim_{\mu\to 0} F_{j,\mu}^{(c)}(\eta,\xi;\rho) = F_j^{(c)}(\eta,\xi;\rho)$, $\lim_{\mu\to 0} F_{j,\mu}^{(s)}(\eta,\xi;\rho) = F_j^{(s)}(\eta,\xi;\rho)$, $(j \geq 0)$, *where* $F_j^{(c)}(\eta,\xi;\rho)$ *and* $F_j^{(s)}(\eta,\xi;\rho)$ *are the new type of Fubini polynomials.*

From Equations (24)–(27), we determine

$$F_{j,\mu}^{(c)}(\eta,\xi;\rho) = \frac{F_{j,\mu}(\eta + i\xi;\rho) + F_{j,\mu}(\eta - i\xi;\rho)}{2}, \tag{28}$$

and

$$F_{j,\mu}^{(s)}(\eta,\xi;\rho) = \frac{F_{j,\mu}(\eta + i\xi;\rho) - F_{j,\mu}(\eta - i\xi;\rho)}{2i}. \tag{29}$$

Theorem 1. *The following result holds true*

$$F_{j,\mu}(\eta + i\xi;\rho) = \sum_{r=0}^{j} \binom{j}{r} (i\xi)_{j-r,\mu} F_{r,\mu}(\eta;\rho)$$
$$= \sum_{r=0}^{j} \binom{j}{r} (\eta + i\xi)_{j-r,\mu} F_{r,\mu}(\rho), \tag{30}$$

and

$$F_{j,\mu}(\eta - i\xi;\rho) = \sum_{r=0}^{j} \binom{j}{r} (-1)^{j-r} (i\xi)_{j-r,\mu} F_{r,\mu}(\eta;\rho)$$
$$= \sum_{r=0}^{j} \binom{j}{r} (-1)^{j-r} (i\xi - \eta)_{j-r,\mu} F_{r,\mu}(\rho), \tag{31}$$

where $(\eta)_{0,\mu} = 1$, $(\eta)_{j,\mu} = \eta(\eta + \mu)\cdots(\eta + \mu(j - 1))$, $(j \geq 1)$.

Proof. From Equation (26), we derive

$$\sum_{j=0}^{\infty} F_{j,\mu}(\eta + i\xi;\rho)\frac{z^j}{j!} = \frac{1}{1-\rho(e_\mu(z)-1)} e_\mu^{\eta}(z) e_\mu^{i\xi}(z)$$
$$= \sum_{r=0}^{\infty} F_{r,\mu}(\eta;\rho)\frac{z^r}{r!} \sum_{j=0}^{\infty} (i\xi)_{j,\mu}\frac{z^j}{j!}$$
$$= \sum_{j=0}^{\infty} \left(\sum_{r=0}^{j} \binom{j}{r} (i\xi)_{j-r,\mu} F_{r,\mu}(\eta;\rho) \right) \frac{z^j}{j!}$$

Similarly, we find

$$\sum_{j=0}^{\infty} F_{j,\mu}(\eta + i\xi;\rho)\frac{z^j}{j!} = \sum_{r=0}^{\infty} F_{r,\mu}(\rho)\frac{z^r}{r!} \sum_{j=0}^{\infty} (\eta + i\xi)_{j,\mu}\frac{z^j}{j!}$$
$$= \sum_{j=0}^{\infty} \left(\sum_{r=0}^{j} \binom{j}{r} (\eta + i\xi)_{j-r,\mu} F_{r,\mu}(\eta;\rho) \right) \frac{z^j}{j!},$$

which implies the asserted result (30). The proof of (31) is similar. $\square$

Theorem 2. *The following result holds true*

$$
\begin{aligned}
F_{j,\mu}^{(c)}(\eta,\xi;\rho) &= \sum_{r=0}^{j} \binom{j}{r} F_{r,\mu}^{(c)}\rho)C_{j-r,\mu}(\eta,\xi) \\
&= \sum_{q=0}^{[\frac{j}{2}]} \sum_{r=2q}^{j} \binom{j}{r} \mu^{r-2q}(-1)^q \xi^{2q} S^{(1)}(r,2q) F_{j-r,\mu}(\eta;\rho),
\end{aligned}
\tag{32}
$$

and

$$
\begin{aligned}
F_{j,\mu}^{(s)}(\eta,\xi;\rho) &= \sum_{r=0}^{j} \binom{j}{r} F_{r,\mu}^{(s)}(\rho)S_{j-r,\mu}(\eta,\xi) \\
&= \sum_{q=0}^{[\frac{j-1}{2}]} \sum_{r=2q+1}^{j} \binom{j}{r} \mu^{r-2q-1}(-1)^q \xi^{2q+1} S^{(1)}(r,2q+1) F_{j-r,\mu}(\eta;\rho).
\end{aligned}
\tag{33}
$$

Proof. From Equations (26) and (16), we find

$$
\begin{aligned}
\sum_{j=0}^{\infty} F_{j,\mu}^{(c)}(\eta,\xi;\rho)\frac{z^j}{j!} &= \frac{1}{1-\rho(e_\mu(z)-1)} e_\mu(\eta z)\cos_\mu(\xi z) \\
&= \left(\sum_{r=0}^{\infty} F_{r,\mu}^{(c)}(\rho)\frac{z^r}{r!} \right) \left(\sum_{j=0}^{\infty} C_{j,\mu}(\eta,\xi)\frac{z^j}{j!} \right) \\
&= \sum_{j=0}^{\infty} \left(\sum_{r=0}^{j} \binom{j}{r} F_{r,\mu}^{(c)}(\rho)C_{j-r,\mu}(\eta,\xi) \right) \frac{z^j}{j!}.
\end{aligned}
\tag{34}
$$

On the other hand, we find

$$
\begin{aligned}
\frac{1}{1-\rho(e_\mu(z)-1)} e_\mu(\eta z)\cos_\mu(\xi z) &= \sum_{j=0}^{\infty} F_{j,\mu}^{(c)}(\eta;\rho)\frac{z^j}{j!} \sum_{r=0}^{\infty} \sum_{q=0}^{[\frac{r}{2}]} \mu^{r-2q}(-1)^q \xi^{2q} S^{(1)}(r,q)\frac{z^r}{r!} \\
&= \sum_{j=0}^{\infty} \left(\sum_{r=0}^{j} \sum_{q=0}^{[\frac{r}{2}]} \binom{j}{r} \mu^{r-2q}(-1)^q \xi^{2q} S^{(1)}(r,2q) F_{j-r,\mu}^{(c)}(\eta;\rho) \right) \frac{z^j}{j!} \\
&= \sum_{j=0}^{\infty} \left(\sum_{q=0}^{[\frac{j}{2}]} \sum_{r=2q}^{n} \binom{j}{r} \mu^{r-2q}(-1)^q \xi^{2q} S^{(1)}(r,2q) F_{j-r,\mu}(\eta;\rho) \right) \frac{z^j}{j!}.
\end{aligned}
\tag{35}
$$

Therefore, by Equations (34) and (35), we obtain (32). The proof of (33) is similar. $\square$

Theorem 3. *The following relation holds true*

$$
C_{j,\mu}(\eta,\xi) = F_{j,\mu}^{(c)}(\eta,\xi;\rho) - \rho \sum_{r=0}^{j} \binom{j}{r} (1)_{r,\mu} F_{j-r,\mu}^{(c)}(\eta,\xi;\rho) + \rho F_{j,\mu}^{(c)}(\eta,\xi;\rho),
\tag{36}
$$

and

$$
S_{j,\mu}(\eta,\xi) = F_{j,\mu}^{(s)}(\eta,\xi;\rho) - \rho \sum_{r=0}^{j} \binom{j}{r} (1)_{r,\mu} F_{j-r,\mu}^{(s)}(\eta,\xi;\rho) + \rho F_{j,\mu}^{(s)}(\eta,\xi;\rho).
\tag{37}
$$

Proof. In view of (16) and (26), we have

$$
e_\mu(\eta z)\cos_\mu(\xi z) = \left[1 - \rho(e_\mu(z)-1) \right] \sum_{j=0}^{\infty} F_{j,\mu}^{(c)}(\eta,\xi;\rho)\frac{z^j}{j!}
$$

$$\sum_{j=0}^{\infty} C_{j,\mu}(\eta,\xi)\frac{z^j}{j!} = \sum_{j=0}^{\infty} F_{j,\mu}^{(c)}(\eta,\xi;\rho)\frac{z^j}{j!} - \rho \sum_{j=0}^{\infty} F_{j,\mu}^{(c)}(\eta,\xi;\rho)\frac{z^j}{j!} \sum_{r=0}^{\infty}(1)_{r,\mu}\frac{z^r}{r!}$$

$$+\rho \sum_{j=0}^{\infty} F_{j,\mu}^{(c)}(\eta,\xi;\rho)\frac{z^j}{j!}$$

$$= \sum_{j=0}^{\infty} F_{j,\mu}^{(c)}(\eta,\xi;\rho)\frac{z^j}{j!} - \rho \sum_{j=0}^{\infty}\sum_{r=0}^{j}\binom{j}{r}(1)_{r,\mu}F_{j-r,\mu}^{(c)}(\eta,\xi;\rho)\frac{z^j}{j!}$$

$$+\rho \sum_{j=0}^{\infty} F_{j,\mu}^{(c)}(\eta,\xi;\rho)\frac{z^j}{j!}.$$

On comparing the coefficients of both sides, we get (36). The proof of (37) is similar. □

3. Main Results

In this section, we derive partial differentiation, recurrence relations, explicit and implicit summation formulae and Stirling numbers of the second kind by using the summation technique series method. We start by the following theorem.

Theorem 4. *For every $j \in \mathbb{N}$, the following equations for partial derivatives hold true:*

$$\frac{\partial}{\partial \eta} F_{j,\mu}^{(c)}(\eta,\xi;\rho) = jF_{j-1,\mu}^{(c)}(\eta,\xi;\rho), \tag{38}$$

$$\frac{\partial}{\partial \xi} F_{j,\mu}^{(c)}(\eta,\xi;\rho) = -jF_{j-1,\mu}^{(s)}(\eta,\xi;\rho), \tag{39}$$

$$\frac{\partial}{\partial \eta} F_{j,\mu}^{(s)}(\eta,\xi;\rho) = jF_{j-1,\mu}^{(s)}(\eta,\xi;\rho), \tag{40}$$

$$\frac{\partial}{\partial \xi} F_{j,\mu}^{(s)}(\eta,\xi;\rho) = jF_{j-1,\mu}^{(c)}(\eta,\xi;\rho). \tag{41}$$

Proof. Using Equation (26), we see

$$\sum_{j=1}^{\infty}\frac{\partial}{\partial \eta} F_{j,\mu}^{(c)}(\eta,\xi;\rho)\frac{z^j}{j!} = \frac{\partial}{\partial \eta}\frac{e_\mu(\eta z)\cos_\mu(\xi z)}{1-\rho(e_\mu(z)-1)} = \sum_{j=0}^{\infty} F_{j,\mu}^{(c)}(\eta,\xi;\rho)\frac{z^{j+1}}{j!}$$

$$= \sum_{j=0}^{\infty} F_{j-1,\mu}^{(c)}(\eta,\xi;\rho)\frac{z^j}{(j-1)!} = \sum_{j=1}^{\infty} nF_{j-1,\mu}^{(c)}(\eta,\xi;\rho)\frac{z^j}{j!},$$

proving (38). Other (39), (40) and (41) can be similarly derived. □

Theorem 5. *For $j \geq 0$, the following formula holds true:*

$$\frac{1}{1-\rho}\sum_{r=0}^{j}\binom{j}{r}F_{r,\mu}\left(\frac{\rho}{1-\rho}\right)C_{j-r,\mu}(\eta,\xi) = \sum_{r=0}^{j}\binom{j}{r}\sum_{q=0}^{\infty}z^q(q)_{r,\mu}C_{j-r,\mu}(\eta,\xi), \tag{42}$$

and

$$\frac{1}{1-\rho}\sum_{r=0}^{j}\binom{j}{r}F_{r,\mu}\left(\frac{\rho}{1-\rho}\right)S_{j-r,\mu}(\eta,\xi) = \sum_{r=0}^{j}\binom{j}{r}\sum_{q=0}^{\infty}z^q(q)_{r,\mu}S_{j-r,\mu}(\eta,\xi). \tag{43}$$

Proof. We begin with the definition (26) and write

$$\sum_{j=0}^{\infty} F_{j,\mu}^{(c)}(\eta,\xi;\rho)\frac{z^j}{j!} = \frac{1}{1-\rho(e_\mu(z)-1)}e_\mu(\eta z)\cos_\mu(\xi z).$$

Let

$$\frac{1}{1-\rho}\left(\frac{1}{1-\frac{\rho}{1-\rho}[e_\mu(z)-1]}\right) = \frac{1}{1-\rho e_\mu(z)} = \sum_{q=0}^{\infty}\rho^q(1+\mu z)^{\frac{q}{\mu}}$$
$$= \sum_{r=0}^{\infty}\left(\sum_{k=0}^{\infty} z^k(k)_{r,\lambda}\right)\frac{t^r}{r!} \tag{44}$$

$$\sum_{j=0}^{\infty} F_{j,\mu}^{(c)}(\eta,\xi;\rho)\frac{z^j}{j!} = \sum_{r=0}^{\infty}\left(\sum_{q=0}^{\infty}\rho^q(q)_{r,\mu}\right)\frac{z^r}{r!}\left(\sum_{j=0}^{\infty} C_{j,\mu}(\eta,\xi)\frac{z^j}{j!}\right)$$
$$= \sum_{j=0}^{\infty}\left(\sum_{r=0}^{j}\binom{j}{r}\sum_{q=0}^{\infty}\rho^q(q)_{r,\mu}C_{j-r,\mu}(\eta,\xi)\right)\frac{z^j}{j!}. \tag{45}$$

Now, we observe that, by (44), we get

$$\frac{1}{1-\rho}\left(\frac{1}{1-\frac{\rho}{1-\rho}(1+\mu z)^{\frac{1}{\mu}}-1}\right) = \frac{1}{1-\rho}\sum_{j=0}^{\infty} F_{j,\mu}\left(\frac{\rho}{1-\rho}\right)\frac{z^j}{j!}.$$

Then, we have

$$\sum_{j=0}^{\infty} F_{j,\mu}^{(c)}(\eta,\xi;\rho)\frac{z^j}{j!} = \frac{1}{1-\rho}\sum_{r=0}^{\infty} F_{r,\mu}\left(\frac{\rho}{1-\rho}\right)\frac{z^r}{r!}\left(\sum_{j=0}^{\infty} C_{j,\mu}(\eta,\xi)\frac{z^j}{j!}\right)$$
$$= \frac{1}{1-\rho}\sum_{j=0}^{\infty}\left(\sum_{r=0}^{j}\binom{j}{r} F_{r,\mu}\left(\frac{\rho}{1-\rho}\right)C_{j-r,\mu}(\eta,\xi)\right)\frac{z^j}{j!}. \tag{46}$$

Therefore, by Equations (45) and (46), we get (42). The proof of (43) is similar. $\square$

Theorem 6. *For $j \geq 0$, the following formula holds true:*

$$C_{j,\mu}(\eta,\xi) = F_{j,\mu}^{(c)}(\eta,\xi;\rho) - \rho F_{j,\mu}^{(c)}(\eta+1,\xi;\rho) + \rho F_{j,\mu}^{(c)}(\eta,\xi;\rho), \tag{47}$$

and

$$S_{j,\mu}(\eta,\xi) = F_{j,\mu}^{(s)}(\eta,\xi;\rho) - \rho F_{j,\mu}^{(s)}(\eta+1,\xi;\rho) + \rho F_{j,\mu}^{(s)}(\eta,\xi;\rho). \tag{48}$$

Proof. We begin with the definition (26) and write

$$e_\mu(\eta z)\cos_\mu(\xi z) = \frac{1-\rho(e_\mu(z)-1)}{1-\rho(e_\mu(z)-1)}e_\mu(\eta z)\cos_\mu(\xi z)$$
$$= \frac{e_\mu(\eta z)\cos_\mu(\xi z)}{1-\rho(e_\mu(z)-1)} - \frac{\rho(e_\mu(z)-1)}{1-\rho(e_\mu(z)-1)}e_\mu(\eta z)\cos_\mu(\xi z).$$

$$\sum_{j=0}^{\infty} C_{j,\mu}(\eta,\xi)\frac{z^j}{j!} = \sum_{j=0}^{\infty}\left[F_{j,\mu}^{(c)}(\eta,\xi;\rho) - \rho F_{j,\mu}^{(c)}(\eta+1,\xi;\rho) + \rho F_{j,\mu}^{(c)}(\eta,\xi;\rho)\right]\frac{z^j}{j!}.$$

Finally, comparing the coefficients of $\frac{z^j}{j!}$, we get (47). The proof of (48) is similar. $\square$

Theorem 7. *For $j \geq 0$ and $\rho_1 \neq \rho_2$, the following formula holds true:*

$$\sum_{q=0}^{j} \binom{j}{q} F_{j-q,\mu}^{(c)}(\eta_1, \xi_1; \rho_1) F_{q,\mu}^{(c)}(\eta_2, \xi_2; \rho_2)$$

$$= \frac{\rho_2 F_{j,\mu}^{(c)}(\eta_1+\eta_2,\xi_1+\xi_2;\rho_2) - \rho_1 F_{n,\mu}^{(c)}(\eta_1+\eta_2,\xi_1+\xi_2;\rho_1)}{\rho_2-\rho_1},$$

(49)

and

$$\sum_{q=0}^{j} \binom{j}{q} F_{j-q,\mu}^{(s)}(\eta_1, \xi_1; \rho_1) F_{q,\mu}^{(s)}(\eta_2, \xi_2; \rho_2)$$

$$= \frac{\rho_2 F_{j,\mu}^{(s)}(\eta_1+\eta_2,\xi_1+\xi_2;\rho_2) - \rho_1 F_{n,\mu}^{(s)}(\eta_1+\eta_2,\xi_1+\xi_2;\rho_1)}{\rho_2-\rho_1}.$$

(50)

Proof. The products of (26) can be written as

$$\sum_{j=0}^{\infty} \sum_{q=0}^{\infty} F_{n,\mu}^{(c)}(\eta_1, \xi_1; \rho_1) \frac{z^j}{j!} F_{q,\mu}^{(c)}(\eta_2, \xi_2; \rho_2) \frac{z^q}{q!} = \frac{e_\mu(\eta_1 z)\cos_\mu(\xi_1 z) e_\mu(\eta_2 z)\cos_\mu(\xi_2 z)}{(1 - \rho_1(e_\mu(z)-1))(1 - \rho_2(e_\mu(z)-1))}$$

$$\sum_{j=0}^{\infty} \left(\sum_{q=0}^{j} \binom{j}{q} F_{j-q,\mu}^{(c)}(\eta_1, \xi_1; \rho_1) F_{q,\mu}^{(c)}(\eta_2, \xi_2; \rho_2) \right) \frac{z^j}{j!}$$

$$= \frac{\rho_2}{\rho_2 - \rho_1} \frac{e_\mu((\eta_1+\eta_2)z)\cos_\mu((\xi_1+\xi_2)z)}{1 - \rho_1(e_\mu(z)-1)} - \frac{\rho_1}{\rho_2 - \rho_1} \frac{e_\mu((\eta_1+\eta_2)z)\cos_\mu((\xi_1+\xi_2)z)}{1 - z_2(e_\lambda(t)-1)}$$

$$= \left(\frac{\rho_2 F_{j,\mu}^{(c)}(\eta_1+\eta_2,\xi_1+\xi_2;\rho_2) - \rho_1 F_{j,\mu}^{(c)}(\eta_1+\eta_2,\xi_1+\xi_2;\rho_1)}{\rho_2 - \rho_1} \right) \frac{z^j}{j!}.$$

By equating the coefficients of $\frac{z^j}{j!}$ on both sides, we get (49). The proof of (50) is similar. $\square$

Theorem 8. *For $j \geq 0$, the following formula holds true:*

$$\rho F_{j,\mu}^{(c)}(\eta + 1, \xi; \rho) = (1+\rho)F_{j,\mu}^{(c)}(\eta, \xi; \rho) - C_{j,\mu}(\eta, \xi),$$

(51)

and

$$\rho F_{j,\mu}^{(s)}(\eta + 1, \xi; \rho) = (1+\rho)F_{j,\mu}^{(s)}(\eta, \xi; \rho) - S_{j,\mu}(\eta, \xi).$$

(52)

Proof. Equation (26), we see

$$\sum_{j=0}^{\infty} \left[F_{j,\mu}^{(c)}(\eta + 1, \xi; \rho) - F_{j,\mu}^{(c)}(\eta, \xi; \rho) \right] \frac{z^j}{j!} = \frac{e_\mu(\eta z)\cos_\mu(\xi z)}{1 - \rho(e_\mu(z)-1)}(e_\mu(z)-1)$$

$$= \frac{1}{\rho} \left[\frac{e_\mu(\eta z)\cos_\mu(\xi z)}{1 - \rho(e_\mu(z)-1)} - e_\mu(\eta z)\cos_\mu(\xi z) \right]$$

$$= \frac{1}{\rho} \sum_{j=0}^{\infty} \left[F_{j,\mu}^{(c)}(x, y; z) - C_{j,\mu}(\eta, \xi) \right] \frac{z^j}{j!}.$$

Comparing the coefficients of $\frac{z^j}{j!}$ on both sides, we obtain (51). The proof of (52) is similar. $\square$

Corollary 1. *The following summation formula holds true*

$$F_{j,\mu}^{(c)}(\eta + 1, \xi; \rho) = \sum_{r=0}^{j} \binom{j}{r} F_{j-r,\mu}^{(c)}(\eta, \xi; \rho)(1)_{r,\mu},$$

and

$$F_{j,\mu}^{(s)}(\eta+1,\xi;\rho) = \sum_{r=0}^{j} \binom{j}{r} F_{j-r,\mu}^{(s)}(\eta,\xi;\rho)(1)_{r,\mu}.$$

Theorem 9. *For $j \geq 0$, then*

$$F_{j,\mu}^{(c)}(\eta+\alpha,\xi;\rho) = \sum_{r=0}^{j} \binom{j}{r} F_{j-r,\mu}^{(c)}(\eta,\xi;\rho)(\alpha)_{r,\mu}, \tag{53}$$

and

$$F_{j,\mu}^{(s)}(\eta+\alpha,\xi;\rho) = \sum_{r=0}^{j} \binom{j}{r} F_{j-r,\mu}^{(s)}(\eta,\xi;\rho)(\alpha)_{r,\mu}. \tag{54}$$

Proof. Replacing η by $\eta+\alpha$ in (26), we have

$$\sum_{j=0}^{\infty} F_{j,\mu}^{(c)}(\eta+\alpha,\xi;\rho)\frac{z^j}{j!} = \frac{1}{1-\rho(e_\mu(z)-1)}e_\mu((\eta+\alpha)z)\cos_\mu(\xi z)$$

$$= \frac{1}{1-\rho(e_\mu(z)-1)}e_\mu(\eta z)\cos_\mu(\xi z)e_\mu(\alpha z)$$

$$= \sum_{j=0}^{\infty} F_{j,\mu}^{(c)}(\eta,\xi;\rho)\frac{z^j}{j!} \sum_{r=0}^{\infty}(\alpha)_{r,\mu}\frac{z^j}{j!}$$

$$= \sum_{j=0}^{\infty} \left(\sum_{r=0}^{j} \binom{j}{r} F_{j-r,\mu}^{(c)}(\eta,\xi;\rho)(\alpha)_{r,\mu} \right) \frac{z^j}{j!}.$$

On comparing the coefficients of z in both sides, we get (53). The proof of (54) is similar. $\square$

Theorem 10. *For $j \geq 0$, the following formula holds true:*

$$F_{j,\mu}^{(c)}(\eta,\xi;\rho) = \sum_{k=0}^{j}\sum_{q=0}^{k} \binom{j}{k} (\eta)_q S_\mu^{(2)}(k,q) F_{j-k,\mu}^{(c)}(0,\xi;\rho), \tag{55}$$

and

$$F_{j,\mu}^{(s)}(\eta,\xi;\rho) = \sum_{k=0}^{j}\sum_{q=0}^{k} \binom{j}{k} (\eta)_q S_\mu^{(2)}(k,q) F_{j-k,\mu}^{(s)}(0,\xi;\rho). \tag{56}$$

Proof. Consider (26), we find

$$\sum_{j=0}^{\infty} F_{j,\mu}^{(c)}(\eta,\xi;\rho)\frac{z^j}{j!} = \frac{1}{1-\rho(e_\mu(z)-1)}[e_\mu(z)-1+1]^\eta \cos_\mu(\xi z)$$

$$= \frac{1}{1-\rho(e_\mu(z)-1)} \sum_{q=0}^{\infty} \binom{\eta}{q} (e_\mu(z)-1)^q \cos_\mu(\xi z)$$

$$= \frac{1}{1-\rho(e_\mu(z)-1)} \cos_\mu(\xi z) \sum_{q=0}^{\infty}(\eta)_q \sum_{k=q}^{\infty} S_\mu^{(2)}(k,q)\frac{z^k}{k!}$$

$$= \sum_{j=0}^{\infty} F_{j,\mu}^{(c)}(0,\xi;\rho)\frac{z^j}{j!} \sum_{k=0}^{\infty} \left(\sum_{q=0}^{k}(\eta)_q S_\mu^{(2)}(k,q) \right) \frac{z^k}{k!}$$

$$= \sum_{j=0}^{\infty} \left(\sum_{k=0}^{j}\sum_{q=0}^{k} \binom{j}{k} (\eta)_q S_\mu^{(2)}(k,q) F_{j-k,\mu}^{(c)}(0,\xi;\rho) \right) \frac{z^j}{j!}.$$

On comparing the coefficients of z in both sides, we get (55). The proof of (56) is similar. $\square$

Theorem 11. *Let $j \geq 0$, then*

$$F_{j,\mu}^{(c)}(\eta, \xi; \rho) = \sum_{r=0}^{j} \binom{j}{r} C_{j-r,\mu}(\eta, \xi) \sum_{k=0}^{r} \rho^k k! S_{2,\mu}(r, k),$$
(57)

and

$$F_{j,\mu}^{(s)}(\eta, \xi; \rho) = \sum_{r=0}^{j} \binom{j}{r} S_{j-r,\mu}(\eta, \xi) \sum_{k=0}^{r} \rho^k k! S_{2,\mu}(r, k).$$
(58)

Proof. Using definition (26), we find

$$\sum_{j=0}^{\infty} F_{j,\mu}^{(c)}(\eta, \xi; \rho) \frac{z^j}{j!} = \frac{1}{1 - \rho(e_\mu(z) - 1)} e_\mu(\eta z) \cos_\mu(\xi z)$$

$$= e_\mu(\eta z) \cos_\mu(\xi z) \sum_{k=0}^{\infty} \rho^k (e_\mu(z) - 1)^k$$

$$= e_\mu(\eta z) \cos_\mu(\xi z) \sum_{k=0}^{\infty} \rho^k k! \sum_{r=k}^{\infty} S_{2,\mu}(r, k) \frac{z^r}{r!}$$

$$= \sum_{j=0}^{\infty} C_{j,\mu}(\eta, \xi) \frac{z^j}{j!} \left(\sum_{r=0}^{\infty} \sum_{k=0}^{r} \rho^k k! S_{2,\mu}(r, k) \frac{z^r}{r!} \right)$$

$$L.H.S = \sum_{j=0}^{\infty} \left(\sum_{r=0}^{j} \binom{j}{r} C_{j-r,\mu}(\eta, \xi) \sum_{k=0}^{r} \rho^k k! S_{2,\mu}(r, k) \right) \frac{z^j}{j!}.$$

Equating the coefficients of $\frac{z^j}{j!}$ in both sides, we get (57). The proof of (58) is similar. $\square$

Theorem 12. *For $j \geq 0$, the following formula holds true:*

$$F_{j,\mu}^{(c)}(\eta + \alpha, \xi; \rho) = \sum_{q=0}^{j} \binom{j}{q} C_{j-q,\mu}(\eta, \xi) \sum_{k=0}^{q} \rho^k k! S_{2,\mu}(q + \alpha, k + \alpha),$$
(59)

and

$$F_{j,\mu}^{(s)}(\eta + \alpha, \xi; \rho) = \sum_{q=0}^{j} \binom{j}{q} S_{j-q,\mu}(\eta, \xi) \sum_{k=0}^{q} \rho^k k! S_{2,\mu}(q + \alpha, k + \alpha).$$
(60)

Proof. Replacing η by $\eta + \alpha$ in (26), we see

$$\sum_{j=0}^{\infty} F_{j,\mu}^{(c)}(\eta + \alpha, \xi; \rho) \frac{z^j}{j!} = \frac{1}{1 - \rho(e_\mu(z) - 1)} e_\mu((\eta + \alpha)z) \cos_\mu(\xi z)$$

$$= e_\mu((\eta + \alpha)z) \cos_\mu(\xi z) e_\mu(rt) \sum_{k=0}^{\infty} \rho^k (e_\mu(z) - 1)^k$$

$$= e_\mu((\eta + \alpha)z) \cos_\mu(\xi z) e_\mu(rt) \sum_{k=0}^{\infty} \rho^k \sum_{q=k}^{\infty} k! S_{2,\mu}(q, k) \frac{z^q}{q!}$$

$$= \sum_{j=0}^{\infty} C_{j,\mu}(\eta, \xi) \frac{z^j}{j!} \sum_{q=0}^{\infty} \rho^k \sum_{k=0}^{q} k! S_{2,\mu}(q + \alpha, k + \alpha) \frac{z^q}{q!}.$$

$$\sum_{j=0}^{\infty} F_{j,\mu}^{(c)}(\eta + \alpha, \xi; \rho) \frac{z^j}{j!}$$

$$= \sum_{n=0}^{\infty} \left(\sum_{q=0}^{j} \binom{j}{q} C_{j-q,\mu}(\eta, \xi) \sum_{k=0}^{q} \rho^k k! S_{2,\mu}(q + \alpha, k + \alpha) \right) \frac{z^j}{j!}.$$

Comparing the coefficients of $\frac{z^j}{j!}$ in both sides, we get (59). The proof of (60) is similar. $\square$

4. Conclusions

In this paper, we study the general properties and identities of the degenerate Fubini polynomials by treating the real and imaginary parts separately, which provide the degenerate cosine Fubini polynomials and degenerate sine Fubini polynomials. These presented results can be applied to any complex Appell type polynomials such as complex Bernoulli and complex Euler polynomials. Furthermore, we show that the degenerate cosine Fubini polynomials and degenerate sine Fubini polynomials can be expressed in terms of the Stirling numbers of the second kind.

Author Contributions: S.K.S., W.A.K., C.S.R. contributed equally to the manuscript and typed, read, and approved final manuscript. All authors have read and agreed to the published version of the manuscript.

Funding: The author would like to thank Deanship of Scientific Research at Majmaah University for supporting this work under Project Number No. R-1441-93.

Acknowledgments: The authors would like to thank the referees for their valuable comments and suggestions.

References

1. Kargin, L. Some formulae for products of Fubini polynomials with applications. *arXiv* **2016**, arXiv:1701.01023v1.
2. Duran, U.; Acikgoz, M. Truncated Fubini polynomials. *Mathematics* **2019**, *7*, 431. [CrossRef]
3. Kim, T.; Kim, D.S.; Jang, G.W. A note on degenerate Fubini polynomials. *Proc. Jangjeon Math. Soc.* **2017**, *20*, 521–531.
4. Kim, D.S.; Kim, T.; Kwon, H.I.; Park, J.W. Two variable higher order Fubini polynomials. *J. Korean Math. Soc.* **2018**, *55*, 975–986. [CrossRef]
5. Kilar, N.; Simsek, Y. A new family of Fubini type numbers and polynomials associated with Apostol-Bernoulli numbers and polynomials. *J. Korean Math. Soc.* **2017**, *54*, 1605–1621.
6. Su, D.D.; He, Y. Some identities for the two variable Fubini polynomial. *Mathematics* **2019**, *7*, 115. [CrossRef]
7. Avram, F.; Taqqu, M.S. Noncentral limit theorems and Appell polynomials. *Ann. Probab.* **1987**, *15*, 767–775. [CrossRef]
8. Tempesta, P. Formal groups, Bernoulli type polynomial and *L*-series. *C. R. Math. Acad. Sci. Paris.* **2007**, *345*, 303–306. [CrossRef]
9. Kim, T.; Ryoo, C.S. Some identities for Euler and Bernoulli polynomials and their zeros. *Axioms* **2018**, *7*, 56. [CrossRef]
10. Carlitz, L. Degenerate Stirling Bernoulli and Eulerian numbers. *Util. Math.* **1979**, *15*, 51–88.
11. Carlitz, L. A degenerate Staud-Clausen theorem. *Arch. Math.* **1956**, *7*, 28–33. [CrossRef]
12. Khan, W.A. A note on degenerate Hermite-poly-Bernoulli numbers and polynomials. *J. Class. Anal.* **2016**, *8*, 65–76. [CrossRef]
13. Haroon, H.; Khan, W.A. Degenerate Bernoulli numbers and polynomials associated with degenerate Hermite polynomials. *Commun. Korean Math. Soc.* **2018**, *33*, 651–669.
14. Kim, D.S.; Dolgy, T.; Komatsu, D.V. Barnes type degenerate Bernoulli polynomials. *Adv. Stud. Contemp. Math.* **2015**, *25*, 121–146.
15. Kim, T. Barnes type multiple degenerate Bernoulli and Euler polynomials. *Appl. Math. Comput.* **2015**, *258*, 556–564. [CrossRef]
16. Kim, D.S.; Kim, T.; Lee, H. A note on degenerate Euler and Bernoulli polynomials of complex variable. *Symmetry* **2019**, *11*, 1168. [CrossRef]

17. Howard, F.T. Degenerate weighted Stirlings numbers. *Discrete Math.* **1985**, *57*, 45–58. [CrossRef]
18. Dil, A.; Kurt, V. Investing geometric and exponential polynomials with Euler-Seidel matrices. *J. Integer Seq.* **2011**, *14*, 1–12.
19. Jamei, M.M.; Beyki, M.R.; Koepf, W. A new type of Euler polynomials and numbers. *Mediterr. J. Math.* **2018**, *15*, 138. [CrossRef]
20. Jamei, M.M.; Beyki, M.R.; Koepf, W. On a Bivariate Kind of Bernoulli Polynomials. Available online: http://www/mathematik.uni-kassel.de/koepf/Publikationen (accessed on 1 February 2020).

 *mathematics*

Article

On 2-Variables Konhauser Matrix Polynomials and Their Fractional Integrals

Ahmed Bakhet [1,2] and Fuli He [1,*]

[1] School of Mathematics and Statistics, Central South University, Changsha 410083, China; kauad_2006@csu.edu.cn

[2] Department of Mathematics, Faculty of Science, Al-Azhar University, Assiut 71524, Egypt

[*] Correspondence: fuli.he@csu.edu.cn or hefuli999@163.com

Received: 28 January 2020; Accepted: 4 February 2020; Published: 10 February 2020

Abstract: In this paper, we first introduce the 2-variables Konhauser matrix polynomials; then, we investigate some properties of these matrix polynomials such as generating matrix relations, integral representations, and finite sum formulae. Finally, we obtain the fractional integrals of the 2-variables Konhauser matrix polynomials.

Keywords: Konhauser matrix polynomial; generating matrix function; integral representation; fractional integral

MSC: 33C25; 33C45; 33D15; 65N35

---

1. Introduction

Special functions play a very important role in analysis, physics, and other applications, and solutions of some differential equations or integrals of some elementary functions can be expressed by special functions. In particular, the family of special polynomials is one of the most useful and applicable family of special functions. The Konhauser polynomials which were first introduced by J.D.E. Konhauser [1] include two classes of polynomials $Y_n^\alpha(x;k)$ and $Z_n^\alpha(x;k)$, where $Y_n^\alpha(x;k)$ are polynomials in x and $Z_n^\alpha(x;k)$ are polynomials in x^k, $\alpha > -1$ and $k \in \mathbb{Z}^+$. Explicit expressions for the polynomials $Z_n^\alpha(x;k)$ are given by

$$Z_n^\alpha(x;k) = \frac{\Gamma(\alpha + kn + 1)}{n!} \sum_{r=0}^{n} (-1)^r \binom{n}{r} \frac{x^{kr}}{\Gamma(\alpha + kr + 1)}, \tag{1}$$

where $\Gamma(\cdot)$ is the classical Gamma function and for the polynomials $Y_n^\alpha(x;k)$, Carlitz [2] subsequently showed that

$$Y_n^\alpha(x;k) = \frac{1}{n!} \sum_{r=0}^{n} \frac{x^r}{r!} \sum_{s=0}^{r} (-1)^s \binom{r}{s} \left(\frac{s+\alpha+1}{k}\right)_n, \tag{2}$$

where $(a)_n$ is Pochhammer's symbol of a as follows:

$$(a)_n = \begin{cases} a(a+1)(a+2)\ldots(a+n-1), & n \geq 1, \\ 1, & n = 0. \end{cases} \tag{3}$$

It is easy to verify that the polynomials $Y_n^\alpha(x;k)$ and $Z_n^\alpha(x;k)$ are biorthogonal with respect to the weight function $w(x) = x^\alpha e^{-x}$ over the interval $(0,\infty)$, which means

$$\int_0^\infty x^\alpha e^{-x} Y_i^\alpha(x;k) Z_j^\alpha(x;k) dx = \frac{\Gamma(kj+\alpha+1)}{j!}\delta_{ij}, \tag{4}$$

where $\alpha > -1$, $k \in \mathbb{Z}^+$ and δ_{ij} is the Kronecker delta.

The Laguerre polynomials $\mathcal{L}_n^\alpha(x)$ are defined as (see, e.g., [3])

$$\mathcal{L}_n^\alpha(x) = \frac{\Gamma(\alpha+n+1)}{\Gamma(n+1)} \sum_{r=0}^n (-1)^r \binom{n}{r} \frac{x^r}{\Gamma(\alpha+r+1)}. \tag{5}$$

For $p,q \in \mathbb{N}$, we can define the general hypergeometric functions of p-numerator and q-denominator by

$${}_pF_q \left[\begin{array}{c} \alpha_1, \alpha_2, \ldots, \alpha_p \\ \beta_1, \beta_2, \ldots, \beta_q \end{array} ; x \right] = \sum_{n=0}^\infty \frac{(\alpha_1)_n (\alpha_2)_n \ldots (\alpha_p)_n}{(\beta_1)_n (\beta_2)_n \ldots (\beta_q)_n} \frac{x^n}{n!}, \tag{6}$$

such that $\beta_j \neq 0, -1, -2, \ldots; j = 1, 2, \ldots, q$. Then, according to [3], we can rewrite $\mathcal{L}_n^\alpha(x)$ as

$$\mathcal{L}_n^\alpha(x) = \frac{(\alpha+1)_n}{n!} {}_1F_1 \left[\begin{array}{c} -n \\ \alpha+1 \end{array} ; x \right]. \tag{7}$$

For $k = 1$, we note that the Konhauser polynomials (1) and (2) reduce to the Laguerre Polynomials $\mathcal{L}_n^\alpha(x)$ and their special cases; when $k = 2$, the case was encountered earlier by Spencer and Fano [4] in certain calculations involving the penetration of gamma rays through matter and was subsequently discussed in [5].

On the other hand, the matrix theory has become pervasive to almost every area of mathematics, especially in orthogonal polynomials and special functions. The special matrix functions appear in the literature related to statistics [6], Lie theory [7], and in connection with the matrix version of Laguerre, Hermite, and Legendre differential equations and the corresponding polynomial families (see, e.g., [8–10]). In the past few years, the extension of the classical Konhauser polynomials to the Konhauser matrix polynomials of one variable has been a subject of intensive studies [11–14]. Recently, many authors (see, e.g., [15–18]) have proposed the generating relations of Konhauser matrix polynomials of one variable from the Lie algebra method point of view and found some properties of Konhauser matrix polynomials of one variable via the Lie algebra technique; they also obtained operational identities for Laguerre–Konhauser-type matrix polynomials and their applications for the matrix framework.

Some studies have been presented on polynomials in two variables such as 2-variables Shivley's matrix polynomials [19], 2-variables Laguerre matrix polynomials [20], 2-variables Hermite generalized matrix polynomials [21–24], 2-variables Gegenbauer matrix polynomials [25], and the second kind of Chebyshev matrix polynomials of two variables [26].

The purpose of the present work is to introduce and study 2-variables Konhauser matrix polynomials and find the hypergeometric matrix function representations; we try to establish some basic properties of these polynomials which include generating matrix functions, finite sum formulae, and integral representations, and we will also discuss the fractional integrals of the 2-variables Konhauser matrix polynomials.

The rest of this paper is structured as follows. In the next section, we give basic definitions and previous results to be used in the following sections. In Section 3, we introduce the definition of 2-variables Konhauser matrix polynomials for parameter matrices A and B and some generating matrix relations involving 2-variables Konhauser matrix polynomials deriving the integral representations. Finally, we provide some results on the fractional integrals of 2-variables Konhauser matrix polynomials in Section 4.

2. Preliminaries

In this section, we give the brief introduction related to Konhauser matrix polynomials and recall some previously known results.

Let $\mathbb{C}^{N \times N}$ be the vector space of N-square matrices with complex entries; for any matrix $A \in \mathbb{C}^{N \times N}$, its spectrum $\sigma(A)$ is the set of all eigenvalues of A,

$$\alpha(A) = \max\{\mathbf{Re}(z) : z \in \sigma(A)\}, \quad \beta(A) = \min\{\mathbf{Re}(z) : z \in \sigma(A)\}. \tag{8}$$

A square matrix $A \in \mathbb{C}^{N \times N}$ is said to be positive stable if and only if $\beta(A) > 0$. Furthermore, the identity matrix and the null matrix or zero matrix in $\mathbb{C}^{N \times N}$ will be symbolized by $\mathbf{I}$ and $\mathbf{0}$, respectively. If $\Phi(z)$ and $\Psi(z)$ are holomorphic functions of the complex variable z, which are defined as an open set Ω of the complex plane and A is a matrix in $\mathbb{C}^{N \times N}$ with $\sigma(A) \subset \Omega$, then, from the properties of the matrix functional calculus [27,28], we have

$$\Phi(A)\Psi(A) = \Psi(A)\Phi(A). \tag{9}$$

Furthermore, if $B \in \mathbb{C}^{N \times N}$ is a matrix for which $\sigma(B) \subset \Omega$ and also if $AB = BA$, then

$$\Phi(A)\Psi(B) = \Psi(B)\Phi(A). \tag{10}$$

Let A be a positive stable matrix in $\mathbb{C}^{N \times N}$. Then, $\Gamma(A)$ is well defined as

$$\Gamma(A) = \int_0^\infty t^{A-I} e^{-t} dt, \tag{11}$$

where $t^{A-I} = \exp((A-I)\ln t)$. Then, the matrix Pochhammer symbol $(A)_n$ of A is denoted as follows (see, e.g., [29–31]):

$$(A)_n = \begin{cases} A(A+I)...(A+(n-1)I) = \Gamma^{-1}(A)\Gamma(A+nI), & n \geq 1, \\ I, & n = 0, \end{cases} \tag{12}$$

The Laguerre matrix polynomials are defined by Jódar et al. [8]

$$L_n^{(A,\lambda)}(x) = \sum_{k=0}^n \frac{(-1)^k \lambda^k}{k!(n-k)!}(A+I)_n[(A+I)_k]^{-1}x^k, \tag{13}$$

where $A \in \mathbb{C}^{N \times N}$ is a matrix such that $-k \notin \sigma(A)$, $\forall k \in \mathbb{Z}^+$, $(A+I)_k$ are given by Equation (12) and λ is a complex number with $\mathbf{Re}(\lambda) > 0$.

For $p, q \in \mathbb{N}$, $1 \leq i \leq p$, $1 \leq j \leq q$, if $A_i, B_j \in \mathbb{C}^{N \times N}$ are matrices such that $B_j + kI$ are invertible for all integers $k \geq 0$, the generalized hypergeometric matrix functions are defined as [32]

$$_pF_q \begin{bmatrix} A_1, A_2, \ldots, A_p \\ B_1, B_2, \ldots, B_q \end{bmatrix} = \sum_{n \geq 0} \frac{(A_1)_n (A_2)_n \ldots (A_p)_n [(B_1)_n]^{-1}[(B_2)_n]^{-1} \ldots ([B_p)_n]^{-1}}{n!} x^n. \tag{14}$$

It follows that, for $\lambda = 1$ in (13), we have

$$L_n^A(x) = \frac{(A+I)_n}{n!} \, _1F_1 \begin{bmatrix} -nI, \\ A+I \end{bmatrix}. \tag{15}$$

For commuting matrices A_i, B_i, C_i, D_i, E_i and F_i in $\mathbb{C}^{N \times N}$, we define the Kampé de Fériet matrix series as [32]

$$
F^{m_1,n_1,l_1}_{m_2,n_2,l_2} \left[\begin{array}{c} A, B, C \\ D, E, F \end{array} ; x, y \right] =
$$

$$
\sum_{m,n \geq 0} \prod_{i=1}^{m_1}(A_i)_{m+n} \prod_{i=1}^{n_1}(B_i)_m \prod_{i=1}^{l_1}(C_i)_n \prod_{i=1}^{m_2}[(D_i)_{m+n}]^{-1} \prod_{i=1}^{n_2}[(E_i)_m]^{-1} \prod_{i=1}^{l_2}[(F_i)_n]^{-1} \frac{x^m y^n}{m! n!}, \tag{16}
$$

where A abbreviates the sequence of matrices $A_1, ..., A_{m_1}$, etc. and $D_i + kI, E_i + kI$ and $F_i + kI$ are invertible for all integers $k \geq 0$.

If $A \in \mathbb{C}^{N \times N}$ is a matrix satisfying the condition

$$
\mathbf{Re}(z) > -1, \quad \forall z \in \sigma(A), \tag{17}
$$

and λ is a complex numbers with $\mathbf{Re}(\lambda) > 0$, we recall the following explicit expression for the Konhauser matrix polynomials (see, e.g., [11])

$$
Z_n^{(A,\lambda)}(x,k) = \frac{\Gamma(A + (kn+1)I)}{n!} \sum_{r=0}^{n}(-1)^r \binom{n}{r} \Gamma^{-1}(A + (kr+1)I)(\lambda x)^{kr}, \tag{18}
$$

and

$$
Y_n^{(A,\lambda)}(x;k) = \frac{1}{n!} \sum_{r=0}^{n} \frac{(\lambda x)^r}{r!} \sum_{s=0}^{r}(-1)^s \binom{r}{s} \left(\frac{A + (s+1)I}{k} \right)_n, \tag{19}
$$

which are biorthogonal with respect to matrix weight function $w(x) = x^A e^{-\lambda x}$ over the interval $(0, \infty)$.

3. 2-Variables Konhauser Matrix Polynomials

In this section, we first introduce the 2-variables Konhauser matrix polynomials with parameter matrices A and B; then, we get the hypergeometric matrix function representations, generating matrix functions, finite summation formulas, and related results for the 2-variables Konhauser matrix polynomials.

Definition 1. *Let* $A, B \in \mathbb{C}^{N \times N}$ *be matrices satisfying the condition* (17). *Then, for* $k, l \in \mathbb{Z}^+$, *the 2-variables Konhauser matrix polynomials* $Z_n^{(A,B,\lambda,\rho)}(x,y,k,l)$ *are defined as follows:*

$$
Z_n^{(A,B,\lambda,\rho)}(x,y,k,l) = \frac{\Gamma(A + (kn+1)I)\Gamma(B + (ln+1)I)}{(n!)^2}
$$

$$
\times \sum_{r=0}^{n} \sum_{s=0}^{n-r} \frac{(-n)_{r+s}}{r! s!} (\lambda x)^{ks}(\rho y)^{lr} \Gamma^{-1}(A + (ks+1)I)\Gamma^{-1}(B + (lr+1)I), \tag{20}
$$

where λ *and* ρ *are complex numbers with* $\mathbf{Re}(\lambda) > 0$ *and* $\mathbf{Re}(\rho) > 0$.

Remark 1. *Furthermore, we note the following special cases of the 2-variables Konhauser matrix polynomials* $Z_n^{(A,B,\lambda,\rho)}(x,y,k,l)$ *as follows:*

i. *Letting* $l = 1$, $B = 0$ *and* $y = 0$ *in* (20), *we get the Konhauser matrix polynomials defined in* (18);

ii. *Letting* $k = l = 1$ *and* $\rho = 1$ *in* (20), *we get the 2-variables analogue of Laguerre's matrix polynomials* $\mathcal{L}_n^{(A,B,\lambda)}(x,y)$ *as follows:*

$$
Z_n^{(A,B,\lambda,1)}(x,y,1,1) = \frac{(A+I)_n(B+I)_n}{(n!)^2} \sum_{r=0}^{n} \sum_{s=0}^{n-r} \frac{(-n)_{r+s}}{r! s!} (\lambda x)^s (y)^r \left[(A+I)_s(B+I)_r \right]^{-1}; \tag{21}
$$

iii. *Letting* $k = l = 1$, $B = 0$ *and* $y = 0$ *in* (20), *we obtain the Laguerre's matrix polynomials* $\mathcal{L}_n^{(A,\lambda)}(x)$ *defined in* (13);

iv. Letting $A = \alpha \in \mathbb{C}^{1\times 1}$ and $B = \beta \in \mathbb{C}^{1\times 1}$ in (20), we find the scaler 2-variables Konhauser polynomials (see, e.g., [33]);

v. Letting $A = \alpha \in \mathbb{C}^{1\times 1}$, and $B = \mathbf{0}$ in (20), we find Konhauser polynomials defined in (1).

3.1. Hypergeometric Representation

Now, by using (16) and (20), we obtain the hypergeometric matrix function representations

$$Z_n^{(A,B,\lambda,\rho)}(x,y,k,l) = \frac{(A+I)_{kn}(B+I)_{ln}}{(n!)^2} F_{k,l}^1 \left[\begin{array}{c} -nI \\ \Delta(k;A+I), \Delta(l;B+I) \end{array} ; (\frac{\lambda x}{k})^k, (\frac{\rho y}{l})^l \right], \qquad (22)$$

where $\Delta(k;A)$ abbreviates the array of k parameters such that

$$\Delta(k;A) = (\frac{A}{k})(\frac{A+I}{k})(\frac{A+2I}{k}) \ldots (\frac{A+(k-1)I}{k}), \quad k \geq 1, \qquad (23)$$

and $F_{k,l}^1$ is defined in (16).

Remark 2. *If $A \in \mathbb{C}^{N\times N}$ is a matrix satisfying the condition (17), letting $B = \mathbf{0}$ and $y = 0$ in (22), we obtain*

$$Z_n^{(A,0,\lambda)}(x,0;k) = \frac{(A+I)_{kn}}{n!} {}_1F_k \left[\begin{array}{c} -nI \\ \Delta(k;A+I) \end{array} ; (\frac{\lambda x}{k})^k \right] = Z_n^{(A,\lambda)}(x;k), \qquad (24)$$

where $Z_n^{(A,\lambda)}(x;k)$ are Konhauser matrix polynomials in [11] and ${}_1F_k$ is hypergeometric matrix function of 1-numerator and k-denominator defined in (14).

Remark 3. *If $A \in \mathbb{C}^{N\times N}$ is a matrix satisfying the condition (17), let $k = 1$, $B = \mathbf{0}$ and $y = 0$ in (22), then we get*

$$Z_n^{(A,\lambda)}(x;1) = \frac{(A+I)_n}{n!} {}_1F_1 \left[\begin{array}{c} -nI, \\ A+I \end{array} ; x \right] = \mathcal{L}_n^A(x), \qquad (25)$$

where $\mathcal{L}_n^A(x)$ are the Laguerre's matrix polynomials defined in (15).

3.2. Generating Matrix Relations for the 2-Variables of Konhauser Matrix Polynomials

Generating matrix relations always play an important role in the study of polynomials, first, we give some generating matrix relations for the 2-variables of Konhauser matrix polynomials as follows:

Theorem 1. *Letting $A, B \in \mathbb{C}^{N\times N}$ be matrices satisfying the condition (17), we obtain the explicit formulae of matrix generating relations for the 2-variables Konhauser matrix polynomials as follows:*

$$\sum_{n-0}^{\infty} Z_n^{(A,B,\lambda,\rho)}(x,y,k,l)[(A+I)_{kn}]^{-1}[(B+I)ln]^{-1}(n!t^n)$$

$$= e^t \, {}_0F_k \left[\begin{array}{c} - \\ \Delta(k;A+I) \end{array} ; (\frac{-\lambda x}{k})^k \right] {}_0F_l \left[\begin{array}{c} - \\ \Delta(p;B+I) \end{array} ; (\frac{-\rho y}{l})^l \right], \qquad (26)$$

where ${}_0F_k$ and ${}_0F_l$ are hypergeometric matrix functions of 0-numerator and k, l-denominator as (14), $\Delta(k;A+I)$ and $\Delta(l;B+I)$ are defined as (23), and the short line "$-$" means that the number of parameters is zero.

Proof. From Equation (20), we have

$$\sum_{n=0}^{\infty} Z_n^{(A,B,\lambda,\rho)}(x,y,k,l)[(A+I)_{kn}]^{-1}[(B+I)ln]^{-1}(n!t^n)$$

$$= \sum_{n=0}^{\infty} n! \sum_{r=0}^{n} \sum_{s=0}^{n-r} \frac{(-n)_{r+s}(\lambda x)^{ks}(\rho y)^{lr}}{r!s!(n!)^2}[(A+I)_{ks}]^{-1}[(B+I)lr]^{-1}t^n \tag{27}$$

$$= \sum_{n=0}^{\infty} \frac{t^n}{n!} \sum_{s=0}^{\infty} \frac{(-1)^s(\lambda x)^{ks}}{s!}[(A+I)_{ks}]^{-1}t^s \sum_{r=0}^{\infty} \frac{(-1)^r(\rho y)^{lr}}{r!}[(B+I)_{lr}]^{-1}t^r,$$

by using

$$(A)_{km} = k^{km}\left(\frac{A}{k}\right)_m\left(\frac{A+I}{k}\right)_m \cdots \left(\frac{A+(k-1)I}{k}\right)_{m'}$$

we get

$$\sum_{n=0}^{\infty} Z_n^{(A,B,\lambda,\rho)}(x,y,k,l)[(A+I)_{kn}]^{-1}[(B+I)ln]^{-1}(n!t^n)$$

$$= \sum_{n=0}^{\infty} \frac{t^n}{n!} \sum_{s=0}^{\infty} \frac{(-1)^s(\lambda x)^{ks}}{k^{ks}s!}\left[\prod_{m=1}^{k}\left(\frac{A+mI}{k}\right)_s\right]^{-1}t^s \sum_{r=0}^{\infty} \frac{(-1)^r(\rho y)^{lr}}{l^{lr}r!}\left[\prod_{n=1}^{l}\left(\frac{B+nI}{l}\right)_r\right]^{-1}t^r \tag{28}$$

$$= e^t\, {}_0F_k\left[\begin{array}{c} - \\ \Delta(k;A+I) \end{array}; \left(\frac{-\lambda x}{k}\right)^k\right]\, {}_0F_l\left[\begin{array}{c} - \\ \Delta(l;B+I) \end{array}; \left(\frac{-\rho y}{l}\right)^l\right].$$

This completes the proof. $\square$

For a matrix E in $\mathbb{C}^{N\times N}$, we can easily obtain the following generating relations for the 2-variables Konhauser matrix polynomial similar to Theorem 1

$$\sum_{n=0}^{\infty}(E)_n[(A+I)_{kn}]^{-1}[(B+I)ln]^{-1}(n!t^n)$$

$$= (1-t)^{-E}F_{k,l}^1\left[\begin{array}{c} -E \\ \Delta(k;A+I),\Delta(l;B+I) \end{array}; \frac{t}{t-1}\left(\frac{\lambda x}{k}\right)^k, \frac{t}{t-1}\left(\frac{\rho y}{l}\right)^l\right], \tag{29}$$

where $F_{k,p}^1$ are defined in Equation (16), $\Delta(k;A+I)$ and $\Delta(l;B+I)$ are defined as Equation (23).

Corollary 1. *Letting $A,B \in \mathbb{C}^{N\times N}$ be matrices satisfying the condition (17), the following generating matrix relations of the 2-variables Konhauser matrix polynomials hold:*

$$\sum_{n=0}^{\infty} Z_n^{(A,B,\lambda,\rho)}(x,y,k,l)\Gamma^{-1}(A+(nk+1)I)\Gamma^{-1}(B+(nl+1)I)(n!t^n)$$

$$= e^t\, \Gamma^{-1}(A+I)\, \Gamma^{-1}(B+I)\, {}_0F_k\left[\begin{array}{c} - \\ \Delta(k;A+I) \end{array}; \left(\frac{-\lambda x}{k}\right)^k\right]\, {}_0F_l\left[\begin{array}{c} - \\ \Delta(l;B+I) \end{array}; \left(\frac{-\rho y}{l}\right)^l\right], \tag{30}$$

where ${}_0F_k$ and ${}_0F_l$ are hypergeometric matrix functions of 0-numerator and k,l-denominator as (14).

Corollary 2. *Letting A, B, and E be matrices in $\mathbb{C}^{N \times N}$ satisfying the condition* (17), *we give explicit formulae of matrix generating relations for the 2-variables Konhauser matrix polynomials as follows:*

$$\sum_{n=0}^{\infty} (E)_n Z_n^{(A,B,\lambda,\rho)}(x,y,k,l)\Gamma^{-1}(A+(nk+1)I)\Gamma^{-1}(B+(nl+1)I)(n!t^n)$$

$$= (1-t)^{-E}\,\Gamma^{-1}(A+I)\,\Gamma^{-1}(B+I)F_{k,l}^1\left[\begin{array}{c} -E \\ \Delta(k;A+I), \Delta(l;B+I) \end{array}; \frac{t}{t-1}\left(\frac{\lambda x}{k}\right)^k, \frac{t}{t-1}\left(\frac{\rho y}{l}\right)^l\right]. \tag{31}$$

Considering the double series,

$$\sum_{n=0}^{\infty}\sum_{m=0}^{\infty} \frac{[(m+n)!]^2}{n!\,m!} Z_n^{(A,B,\lambda,\rho)}(x,y,k,l)[(A+I)_{k(m+n)}]^{-1}[(B+I)l(m+n)]^{-1}\sigma^m \tau^n$$

$$= \sum_{n=0}^{\infty} n! Z_n^{(A,B,\lambda,\rho)}(x,y,k,l)\tau^n[(A+I)_{kn}]^{-1}[(B+I)ln]^{-1}\,{}_1F_0\left[\begin{array}{c} -nI \\ - \end{array}; \frac{-\sigma}{\tau}\right] \tag{32}$$

$$= \sum_{n=0}^{\infty} n! Z_n^{(A,B,\lambda,\rho)}(x,y,k,l)[(A+I)_{kn}]^{-1}[(B+I)ln]^{-1}(\sigma+\tau)^n.$$

Now, by making use of Theorem 1, we find

$$\sum_{n=0}^{\infty}\sum_{m=0}^{\infty} \frac{[(m+n)!]^2}{n!\,m!} Z_n^{(A,B,\lambda,\rho)}(x,y,k,l)[(A+I)_{k(m+n)}]^{-1}[(B+I)l(m+n)]^{-1}\sigma^m \tau^n$$

$$= e^{\sigma+\tau}\,{}_0F_k\left[\begin{array}{c} - \\ \Delta(k;A+I) \end{array}; \left(\frac{-\lambda x}{k}\right)^k(\sigma+\tau)\right]{}_0F_l\left[\begin{array}{c} - \\ \Delta(l;B+I) \end{array}; \left(\frac{-\rho y}{l}\right)^l(\sigma+\tau)\right]. \tag{33}$$

Here, Equation (33) may be regarded as a double generating matrix relations for (20).

Remark 4. *For A in $\mathbb{C}^{N \times N}$, letting $k = 1$, $B = 0$ and $y = 0$ in* (33), *we have*

$$\sum_{n=0}^{\infty} \binom{m+n}{n}[(A+I)_{(m+n)}]^{-1}\mathcal{L}_{m+n}^A(x)\,t^n$$

$$= \sum_{n=m}^{\infty} \frac{(-1)^n m![(A+I)_n]^{-1}x^n}{(n-m)!n!}\,{}_1F_1\left[\begin{array}{c} -(n+1)I, \\ (n-m+1)I \end{array}; t\right]$$

$$= \sum_{n=m}^{\infty}\sum_{j=0}^{\infty} \frac{(-x)^n n! t^{n-m}[(A+I)_n]^{-1}(n+1)_j t^j}{m!(n-m)!n!(n-m+1)j!} \tag{34}$$

$$= \sum_{n=0}^{\infty}\sum_{j=0}^{\infty} \frac{(-x)^n(n+1)_j(A+I)_n]^{-1}t^j}{(1)_j n!j!},$$

we find generating matrix relations of the Laguerre's matrix polynomials.

3.3. Some Properties of the 2-Variables Konhauser Matrix Polynomials

For the finite sum property of the 2-variables Konhauser matrix polynomials $Z_n^{(A,B,\lambda,\rho)}(x,y,k,l)$, we get the generating relations together as follows:

$$e^t\,{}_0F_k\left[\begin{array}{c} - \\ \Delta(k;A+I) \end{array}; \left(\frac{-\lambda xw}{k}\right)^k t\right]{}_0F_l\left[\begin{array}{c} - \\ \Delta(l;B+I) \end{array}; \left(\frac{-\rho yw}{l}\right)^l t\right]$$

$$= e^{(1-w^k)t}e^{w^k t}\,{}_0F_k\left[\begin{array}{c} - \\ \Delta(k;A+I) \end{array}; \left(\frac{-\lambda xw}{k}\right)^k t\right]{}_0F_l\left[\begin{array}{c} - \\ \Delta(l;B+I) \end{array}; \left(\frac{-\rho yw}{l}\right)^l t\right], \tag{35}$$

and

$$\sum_{r=0}^{n} Z_n^{(A,B,\lambda,\rho)}(xw, yw, k, k)[(A+I)_{kn}]^{-1}[(B+I)_{kn}]^{-1}t^n n!$$
$$= \left(\sum_{n=0}^{\infty} \frac{1 - w^{kn} t^n}{n!} \right) \left(\sum_{n=0}^{\infty} Z_n^{(A,B,\lambda,\rho)}(x, y, k, k) w^{kn}[(A+I)_{kn}]^{-1}[(B+I)_{kn}]^{-1}t^n n! \right). \tag{36}$$

By comparing the coefficients of t^n on both sides, we have

$$Z_n^{(A,B,\lambda,\rho)}(xw, yw, k, k)$$
$$= \sum_{r=0}^{n} \frac{r! w^{kr}(1 - w^k)^{n-r}}{n!(n-r)!}[(A+I)_{kr}]^{-1}[(B+I)_{kr}]^{-1}(A+I)_{kn}(B+I)_{kn} Z_n^{(A,B,\lambda,\rho)}(x, y, k, k). \tag{37}$$

The integral representations for the 2-variables Konhauser matrix polynomials are derived in the following theorem.

Theorem 2. *Letting $A, B \in \mathbb{C}^{N \times N}$ be matrices satisfying the condition (17), and, if $\left| \frac{t}{\lambda x} \right| < 1, \left| \frac{v}{\rho y} \right| < 1$, then we have the integral representation of the 2-variables Konhauser matrix polynomials $Z_n^{(A,B,\lambda,\rho)}(x, y, k, l)$ as follows:*

$$Z_n^{(A,B,\lambda,\rho)}(x, y, k, l) = \frac{\Gamma(A + (kn+1)I)\Gamma(B + (ln+1)I)}{(n!)^2(2\pi i)^2}$$
$$\times \int_{c_1} \int_{c_2} \left(t^k v^l - (\lambda x)^k v^l - (\rho y)^k t^l \right)^n e^{t+v} \, t^{-(A+(kn+1)I)} \, v^{-(B+(ln+1)I)} dt dv, \tag{38}$$

where c_1, c_2 are the paths around the origin in the positive direction, beginning at and returning to positive infinity with respect for the branch cut along the positive real axis.

Proof. The right side of the above formulae are deformed into

$$\frac{\Gamma(A + (kn+1)I)\Gamma(B + (ln+1)I)}{(n!)^2} \sum_{r=0}^{n} \sum_{s=0}^{n-r} \frac{(-n)_{r+s}(\lambda x)^{kr}(\rho y)^{lr}}{r! s!}$$
$$\times \frac{1}{2\pi i} \int_{c_1} t^{-(A+(ks+1)I)} e^t dt \times \frac{1}{2\pi i} \int_{c_2} v^{-(B+(lr+1)I)} e^v dv, \tag{39}$$

and using the integral representation of the reciprocal Gamma function, which are given in [34]

$$\frac{1}{\Gamma(z)} = \frac{1}{2\pi i} \int_c e^t t^{-z} dt, \tag{40}$$

where c is the path around the origin in the positive direction, beginning at and returning to positive infinity with respect for the branch cut along the positive real axis. Thus, from Equation (40), we obtain the following integral matrix functional

$$\Gamma^{-1}(A + (kn+1)I) = \frac{1}{2\pi i} \int_{c_1} e^t t^{-(A+(kn+1)I)} dt. \tag{41}$$

By Equation (41), we can transfer (39) to

$$\frac{\Gamma(A + (kn+1)I)\Gamma(B + (ln+1)I)}{(n!)^2} \times$$
$$\sum_{r=0}^{n} \sum_{s=0}^{n-r} \frac{(-n)_{r+s}(\lambda x)^{kr}(\rho y)^{lr}}{r! s!} \Gamma^{-1}(A + (ks+1)I)\Gamma^{-1}(B + (kr+1)I) \tag{42}$$
$$= Z_n^{(A,B,\lambda,\rho)}(x, y, k, l).$$

This completes the proof of the theorem. $\quad\square$

4. Fractional Integrals of the 2-Variable Konhauser Matrix Polynomials

In this section, we study the fractional integrals of the Konhauser matrix polynomials of one and two variables. The fractional integrals of Riemann–Liouville operators of order μ and $x > 0$ are given by (see [35,36])

$$(\mathbf{I}_a^\mu f)(x) = \frac{1}{\Gamma(\mu)} \int_a^x (x-t)^{\mu-1} f(t) dt, \quad \mathbf{Re}(\mu) > 0. \tag{43}$$

Recently, the authors (see, e.g., [28]) introduced the fractional integrals with matrix parameters as follows: suppose $A \in \mathbb{C}^{N \times N}$ is a positive stable matrix and $\mu \in \mathbb{C}$ is a complex number satisfying the condition $\mathbf{Re}(\mu) > 0$. Then, the Riemann–Liouville fractional integrals with matrix parameters of order μ are defined by

$$\mathbf{I}^\mu(x^A) = \frac{1}{\Gamma(\mu)} \int_0^x (x-t)^{\mu-1} t^A dt. \tag{44}$$

Lemma 1. *Supposing that $A \in \mathbb{C}^{N \times N}$ is a positive stable matrix and $\mu \in \mathbb{C}$ is a complex number satisfying the condition $\mathbf{Re}(\mu) > 0$, then the Riemann–Liouville fractional integrals with matrix parameters of order μ are defined and we have (see, e.g., [28])*

$$\mathbf{I}^\mu(x^{A-I}) = \Gamma(A)\Gamma^{-1}(A + \mu I) x^{A+(\mu-1)I}. \tag{45}$$

Theorem 3. *If $A \in \mathbb{C}^{N \times N}$ is a matrix satisfying the condition (17), then the Riemann–Liouville fractional integrals of Konhauser matrix polynomials of one variable are as follows:*

$$\mathbf{I}^\mu\left[(\lambda x)^A Z_n^{(A,\lambda)}(x,k)\right] = \Gamma^{-1}(A + (kn + \mu + 1)I)\Gamma(A + (kn+1)I)(\lambda x)^{A+\mu I} Z_n^{(A+\mu I,\lambda)}(x,k), \tag{46}$$

where λ is a complex numbers with $\mathbf{Re}(\lambda) > 0$, and $k \in \mathbb{Z}^+$.

Proof. From Equation (44), we find

$$\mathbf{I}^\mu\left[(\lambda x)^A Z_n^{(A,\lambda)}(x,k)\right] = \int_0^x \frac{(\lambda(x-t))^{\mu-1}}{\Gamma(\mu)} t^A Z_n^{(A,\lambda)}(t,k) dt$$

$$= \frac{\Gamma(A + (kn+1)I)}{\Gamma(\mu)} \sum_{r=0}^n \frac{(-1)^r}{r!(n-r)!} \Gamma^{-1}(A + (kr+1)I) \int_0^x (\lambda x)^{A+krI} (\lambda(x-t))^{\mu-1} dt \tag{47}$$

$$= \Gamma(A + (kn+1)I) \sum_{r=0}^n \frac{(-1)^r}{r!(n-r)!} (\lambda x)^{A+(kr+\mu)I} \Gamma^{-1}(A + (kr+\mu+1)I),$$

and we can write

$$\mathbf{I}^\mu\left[(\lambda x)^A Z_n^{(A,\lambda)}(x,k)\right] = \Gamma^{-1}(A + (kn+\mu+1)I)\Gamma(A + (kn+1)I)(\lambda x)^{A+\mu I} Z_n^{(A+\mu I,\lambda)}(x,k). \tag{48}$$

$\square$

The 2-variables analogue of Riemann–Liouville fractional integrals $\mathbf{I}^{\nu,\mu}$ may be defined as follows

Definition 2. *Letting $A, B \in \mathbb{C}^{N \times N}$ be positive stable matrices, if $\mathbf{Re}(\nu) > 0$ and $\mathbf{Re}(\mu) > 0$, then the 2-variables Riemann–Liouville fractional integrals of orders ν, μ can be defined as follows:*

$$\mathbf{I}^{\nu,\mu}\left[x^A y^B\right] = \frac{1}{\Gamma(\nu)\Gamma(\mu)} \int_0^x \int_0^y (x-u)^{\nu-1}(y-v)^{\mu-1} u^A v^B du dv. \tag{49}$$

Theorem 4. *Letting $A, B \in \mathbb{C}^{N \times N}$ be matrices satisfying the condition* (17), $\mathbf{Re}(\lambda) > 0$, $\mathbf{Re}(\rho) > 0$, *then, for the Riemann–Liouville fractional integral of a 2-variables Konhauser matrix polynomial, we have the following:*

$$\mathbf{I}^{\nu,\mu}\left[(\lambda x)^A (\rho y)^B Z_n^{(A,B,\lambda,\rho)}(x,y,k,l)\right]$$
$$= \Gamma^{-1}(A + (kn + \nu + 1)I)\Gamma^{-1}(B + (ln + \mu + 1)I)\Gamma(A + (kn + 1)I) \tag{50}$$
$$\Gamma(B + (ln + 1)I)(\lambda x)^{A+\nu I}(\rho y)^{B+\mu I} Z_n^{(A+\nu I, B+\mu I, \lambda, \rho)}(x,y,k,l),$$

where λ and ρ are complex numbers and $k, l \in \mathbb{Z}^+$.

Proof. By using Equation (49), we obtain

$$\mathbf{I}^{\nu,\mu}\left[(\lambda x)^A (\rho y)^B Z_n^{(A,B,\lambda,\rho)}(x,y,k,l)\right] = \frac{1}{\Gamma(\nu)\Gamma(\mu)}$$
$$\times \int_0^x \int_0^y \left(\lambda(x - u)\right)^{\nu-1}\left(\rho(y - v)\right)^{\mu-1}(\lambda u)^A(\rho v)^B Z_n^{(A,B,\lambda,\rho)}(u,v,k,l)\,du\,dv. \tag{51}$$

By putting $u = xt$ and $v = yw$, we get

$$\mathbf{I}^{\nu,\mu}\left[(\lambda x)^A (\rho y)^B Z_n^{(A,B,\lambda,\rho)}(x,y,k,l)\right] = \frac{(\lambda x)^{A+\nu I}(\rho y)^{B+\mu I}}{\Gamma(\nu)\Gamma(\mu)}$$
$$\times \int_0^1 \int_0^1 (\lambda t)^A(\rho w)^B \left(\lambda(1 - t)\right)^{\nu-1}\left(\rho(1 - w)\right)^{\mu-1} Z_n^{(A,B,\lambda,\rho)}(xt, yw, k, l)\,dt\,dw, \tag{52}$$

from definition (20), we have

$$\mathbf{I}^{\nu,\mu}\left[(\lambda x)^A (\rho y)^B Z_n^{(A,B,\lambda,\rho)}(x,y,k,l)\right]$$
$$= \frac{\Gamma(A + (kn + 1)I)\Gamma(B + (ln + 1)I)(\lambda x)^{A+\nu I}(\rho y)^{B+\mu I}}{(n!)^2 \Gamma(\nu)\Gamma(\mu)}$$
$$\sum_{r=0}^n \sum_{s=0}^{n-r} \frac{(-n)_{r+s}\,(\lambda x)^{ks}(\rho y)^{lr}}{r!s!}\Gamma^{-1}(A + (ks + 1)I)\Gamma^{-1}(B + (lr + 1)I). \tag{53}$$
$$\times \int_0^1 (\lambda t)^{A+ksI}\left(\lambda(1 - t)\right)^{\nu-1}dt \int_0^1 (\rho w)^{B+lrI}\left(\rho(1 - w)\right)^{\mu-1}dw,$$

and

$$\mathbf{I}^{\nu,\mu}\left[(\lambda x)^A (\rho y)^B Z_n^{(A,B,\lambda,\rho)}(x,y,k,l)\right]$$
$$= \frac{\Gamma(A + (kn + 1)I)\Gamma(B + (ln + 1)I)(\lambda x)^{A+\nu I}(\rho y)^{B+\mu I}}{(n!)^2 \Gamma(\nu)\Gamma(\mu)}$$
$$\sum_{r=0}^n \sum_{s=0}^{n-r} \frac{(-n)_{r+s}\,(\lambda x)^{ks}(\rho y)^{lr}}{r!s!}\Gamma^{-1}(A + (ks + \nu + 1)I)\Gamma^{-1}(B + (lr + \mu + 1)I). \tag{54}$$

We thus arrive at

$$\mathbf{I}^{\nu,\mu}\left[(\lambda x)^A (\rho y)^B Z_n^{(A,B,\lambda,\rho)}(x,y,k,l)\right]$$
$$= \Gamma^{-1}(A + (kn + \nu + 1)I)\Gamma^{-1}(B + (ln + \mu + 1)I)\Gamma(A + (kn + 1)I) \tag{55}$$
$$\Gamma(B + (ln + 1)I)(\lambda x)^{A+\nu I}(\rho y)^{B+\mu I} Z_n^{(A+\nu I, B+\mu I, \lambda, \rho)}(x,y,k,l).$$

This completes the proof of Theorem 4. $\square$

Author Contributions: All authors contributed equally and all authors have read and agreed to the published version of the manuscript.

Funding: This research is supported by the National Natural Science Foundation of China (11601525).

Conflicts of Interest: The authors declare no conflict of interest.

References

1. Konhauser, J.D.E. Biorthogonal polynomials suggested by the Laguerre polynomials. *Pac. J. Math.* **1967**, *21*, 303–314. [CrossRef]
2. Carlitz, L. A note on Certain Biorthogonal Polynomials. *Pac. J. Math.* **1968**, *24*, 425–430. [CrossRef]
3. Rainville, E.D. *Special Functions*; Macmillan: New York, NY, USA; Chelsea Publishing Co.: Bronx, NY, USA, 1997.
4. Spencer, L.; Fano, U. Penetration and diffusion of X-rays, Calculation of spatial distribution by polynomial expansion. *J. Res. Nat. Bur. Standards* **1951**, *46*, 446–461. [CrossRef]
5. Preiser, S. An investigation of biorthogonal Polynomials derivable from ordinary differential equation of the third order. *J. Math. Anal. Appl.* **1962**, *4*, 38–64. [CrossRef]
6. Constantine, G.; Muirhead, R.J. Partial differential equations for hypergeometric functions of two argument matrix. *J. Multivariate Anal.* **1972**, *3*, 332–338. [CrossRef]
7. James, T. Special Functions of Matrix and Single Argument in Statistics. In *Theory and Applications of Special functions*; Askey, R.A., Ed.; Academic Press: New York, NY, USA, 1975; pp. 497–520.
8. Jódar, L.; Company, R.; Navarro, E. Laguerre matrix polynomials and systems of second-order differential equations. *Appl. Numer. Math.* **1994**, *15*, 53–63. [CrossRef]
9. Jódar, L.; Company, R.; Ponsoda, E. Orthogonal matrix polynomials and systems of second order differential equations. *Diff. Equ. Dyn. Syst.* **1995**, *3*, 269–288.
10. Jódar, L.; Company, R. Hermite matrix polynomials and second order differential equations. *Approx. Theory Appl.* **1996**, *12*, 20–30.
11. Varma, S.; Çekim, B.; Taşdelen, F. Konhauser matrix polynomials. *Ars. Combin.* **2011**, *100*, 193–204.
12. Erkuş-Duman, E.; Çekim, B. New generating functions for Konhauser matrix polynomials. *Commun. Fac. Sci. Univ. Ank. Ser. A1 Math. Stat.* **2014**, *63*, 35–41.
13. Varma, S.; Taşdelen, F. Some properties of Konhauser matrix polynomials. *Gazi Univ. J. Sci.* **2016**, *29*, 703–709.
14. Shehata, A. Some relations on Konhauser matrix polynomials. *Miskolc Math. Notes.* **2016**, *17*, 605–633. [CrossRef]
15. Shehata, A. Certain generating relations of Konhauser matrix polynomials from the view point of Lie algebra method. *Univ. Politech. Bucharest Sci. Bull. Ser. A Appl. Math. Phys.* **2017**, *79*, 123–136.
16. Shehata, A. Certain properties of Konhauser matrix polynomials via Lie algebra technique. *Boletín de la Sociedad Matemática Mexicana* **2019**. [CrossRef]
17. Shehata, A. A note on Konhauser matrix polynomials. *Palestine J. Math.* **2020**, *9*, 549–556.
18. Bin-Saad, M.G.; Mohsen, F.B.F. Quasi-monomiality and operational identities for Laguerre-Konhauser-type matrix polynomials and their applications. *Acta Comment. Univ. Tartuensis Math.* **2018**, *22*, 13–22. [CrossRef]
19. He, F.; Bakhet, A.; Hidan, M.; Abdalla, M. Two Variables Shivley's Matrix Polynomials. *Symmetry* **2019**, *11*, 151. [CrossRef]
20. Khan, S.; Hassan, N.A.M. 2-variables Laguerre matrix polynomials and Lie-algebraic techniques. *J. Phys. A Math. Theor.* **2010**, *43*, 235204. [CrossRef]
21. Batahan, R.S. A new extension of Hermite matrix polynomials and its applications. *Lin. Algebra. Appl.* **2006**, *419*, 82–92. [CrossRef]
22. Srivastava, H.M.; Khan, W.A.; Haroon, H. Some expansions for a class of generalized Humbert matrix polynomials. *Revista de la Real Academia de Ciencias Exactas Físicas y Naturales Serie A Matemáticas* **2019**, *113*, 3619–3634. [CrossRef]
23. Khan, S.; Raza, N. 2-variable generalized Hermite matrix polynomials and Lie algebra representation. *Rep. Math. Phys.* **2010**, *66*, 159–174. [CrossRef]
24. Khan, S.; Al-Gonah, A. Multi-variable Hermite matrix polynomials: Properties and applications. *J. Math. Anal. Appl.* **2014**, *412*, 222–235. [CrossRef]
25. Kahmmash, G.S. A study of a two variables Gegenbauer matrix polynomials and second order matrix partial differential equations. *Int. J. Math. Anal.* **2008**, *2*, 807–821.

26. Kargin, L.; Kurt,V. Chebyshev-type matrix polynomials and integral transforms. *Hacett. J. Math. Stat.* **2015**, *44*, 341–350. [CrossRef]

27. Abdalla, M. Special matrix functions: Characteristics, achievements and future directions. *Linear Multilinear Algebra* **2020**, *68*, 1–28. [CrossRef]

28. Bakhet, A.; Jiao, Y.; He, F. On the Wright hypergeometric matrix functions and their fractional calculus. *Integr. Transf. Spec. Funct.* **2019**, *30*, 138–156. [CrossRef]

29. Cortés, J.C.; Jódar, L.; Sols, F.J.; Ku Carrillo, R. Infinite matrix products and the representation of the gamma matrix function. *Abstr. Appl. Anal.* **2015**, *3*, 1–8. [CrossRef]

30. Jódar, L.; Cortés, J.C. Some properties of Gamma and Beta matrix functions. *Appl. Math. Lett.* **1998**, *11*, 89–93. [CrossRef]

31. Jódar, L.; Cortés, J.C. On the hypergeometric matrix function. *J. Comp. Appl. Math.* **1998**, *99*, 205–217. [CrossRef]

32. Dwivedi, R.; Sahai, V. On the hypergeometric matrix functions of two variables. *Linear Multilinear Algebra* **2018**, *66*, 1819–1837. [CrossRef]

33. Khan, M.A.; Ahmed, K. On a two variables analogue of Konhauser's biorthogonal polynomial $Z^{\alpha}(x; k)$. *Far East J. Math. Sci.* **1999**, *1*, 225–240.

34. Lebedev, N.N. *Special Functions and Their Applications*; Dover Publications Inc.: New York, NY, USA, 1972.

35. Samko, S.G.; Kilbas, A.A.; Marichev, O.I. *Fractional Integrals and Derivatives: Theory and Applications*; Gordon & Breach: Yverdon, Switzerland, 1993.

36. Kilbas, A.A.; Srivastava, H.M.; Trujillo, J.J. *A Theory and Applications of Fractional Differential Equations*; North-Holland Mathematical Studies, 204; Elsevier: Amsterdam, The Netherlands, 2006.

 *mathematics*

Article

Fractional Supersymmetric Hermite Polynomials

Fethi Bouzeffour [1,*] and Wissem Jedidi [2,3]

[1] Department of Mathematics, College of Sciences, King Saud University, P. O. Box 2455, Riyadh 11451, Saudi Arabia
[2] Department of Statistics & OR, King Saud University, P.O. Box 2455, Riyadh 11451, Saudi Arabia; wissem.jedidi@fst.utm.tn
[3] Faculté des Sciences de Tunis, LR11ES11 Laboratoire d'Analyse Mathématiques et Applications, Université de Tunis El Manar, Tunis 2092, Tunisia
* Correspondence: fbouzaffour@ksu.edu.sa

Received: 12 December 2019; Accepted: 31 January 2020; Published: 5 February 2020

Abstract: We provide a realization of fractional supersymmetry quantum mechanics of order r, where the Hamiltonian and the supercharges involve the fractional Dunkl transform as a Klein type operator. We construct several classes of functions satisfying certain orthogonality relations. These functions can be expressed in terms of the associated Laguerre orthogonal polynomials and have shown that their zeros are the eigenvalues of the Hermitian supercharge. We call them the supersymmetric generalized Hermite polynomials.

Keywords: orthogonal polynomials; difference-differential operator; supersymmetry

---

1. Introduction

Supersymmetry relates bosons and fermions on the basis of $\mathbb{Z}_2$-graded superalgebras [1,2], where the fermionic set is realized in terms of matrices of finite dimension or in terms of Grassmann variables [3]. The supersymmetric quantum mechanics (SUSYQM), introduced by Witten [2], may be generated by three operators Q_-, Q_+ and H satisfying

$$Q_\pm^2 = 0, \quad [Q_\pm, H] = 0, \quad \{Q_-, Q_+\} = H. \tag{1}$$

Superalgebra (1) corresponds to the case $N = 2$ supersymmetry. The usual construction of Witten's supersymmetric quantum mechanics with the superalgebra (1) is performed by introduction of fermion degrees of freedom (realized in a matrix form, or in terms of Grassmann variables) which commute with bosonic degrees of freedom. Another realization of supersymmetric quantum mechanics, called minimally bosonized supersymmetric quantum [1,4,5], is built by taking the supercharge as the following Dunkl-type operator:

$$Q = \partial_x R + v(x),$$

where $v(x)$ is a superpotential.

The fractional supersymmetric quantum mechanics of order r (FSUYQM) are an extension of the ordinary supersymmetric quantum mechanics for which the $\mathbb{Z}_2$-graded superalgebras are replaced by a $\mathbb{Z}_r$-graded superalgberas [3,6,7]. The framework of the fractional supersymmetric quantum mechanics has been shown to be quite fruitful. Amongst many works, we may quote the deformed Heisenberg algebra introduced in connection with parafermionic and parabosonic systems [3,4], the C_λ-extended oscillator algebra developed in the framework of parasupersymmetric quantum mechanics [8], and the generalized Weyl–Heisenberg algebra W_k related to $\mathbb{Z}_k$-graded supersymmetric quantum mechanics [3].

Note that the construction of fractional supersymmetric quantum mechanics without employment of fermions and parafermions degrees of freedom was started in [4,9,10]. In particular, the idea of realization of fractional supersymmetry in the form as it was presented in [3,8] was initially proposed in [4] and also in [9]. In this work, we develop a fractional supersymmetric quantum of order r without parafermonic degrees of freedom. We essentially use a difference-differential operators generated from a special case of the well known fractional Dunkl transform. We then investigate the characteristics of the (r)-scheme.

The paper is organized as follows. In Section 2, we discuss some of basic properties of the fractional Dunkl transform and we define the generalized Klein operator. In Section 3, we present a realization of the fractional supersymmetric quantum mechanics and we construct a basis involving the generalized Hermite functions that diagonalize the Hamiltonian. In Section 4, we define the associated generalized Hermite polynomials and we provide its weight function and we show that the eigenvalues of the supercharge are the zeros of the associated generalized Hermite polynomials.

2. Preliminaries

Recall that the fractional Dunkl transform on the real line, introduced in [11,12], is both an extension of the fractional Hankel transform and the Fourier transform. For $0 < |\alpha| < \pi$, the fractional Dunkl transform is defined by:

$$\mathcal{F}_\nu^\alpha f(t) = \frac{e^{i(\nu+1/2)(\tilde{\alpha}\pi/2-\alpha)}}{(2|\sin(\alpha)|)^{\nu+1/2}\Gamma(\nu+1/2)} \int_{\mathbb{R}} e^{-i\frac{t^2+x^2}{2\tan\alpha}} \mathcal{E}_\nu\left(\frac{itx}{\sin\alpha}\right) f(x)\,|x|^{2\nu}dx,$$

where

$$\tilde{\alpha} = \text{sgn}(\sin(\alpha))$$

and

$$\begin{aligned}
\mathcal{E}_\nu(x) &:= \mathcal{J}_{\nu-1/2}(ix) + \frac{x}{2\nu+1}\mathcal{J}_{\nu+1/2}(ix), \\
\mathcal{J}_\nu(x) &:= \Gamma(\nu+1)\,(2/x)^\nu J_\nu(x).
\end{aligned}$$

Notice that $J_\nu(x)$ is the standard Bessel function ([13] Ch. 10) and $\Gamma(x)$ is the Gamma function. It is well known that, for $\nu > 0$, the function $\mathcal{E}_\nu(\lambda x)$ is the unique analytic solution of the following system that can be found in [14]:

$$\begin{cases} Y_\nu \mathcal{E}_\nu(\lambda x) = i\lambda\,\mathcal{E}_\nu(\lambda x), \\ \mathcal{E}_\nu(0) = 1, \end{cases} \tag{2}$$

where Y_ν is the Dunkl operator related to root system A_1 (see ([14] Definition 4.4.2))), which is a differential-difference operator, depending on a parameter $\nu \in \mathbb{R}$ and acting on $C^\infty(\mathbb{R})$ as:

$$Y_\nu := \frac{d}{dx} + \frac{\nu}{x}(1-R), \tag{3}$$

where R is the Klein operator :

$$(Rf)(x) = f(-x). \tag{4}$$

The operator Y_ν is also related by a simple similarity transformation to the Yang–Dunkl operator used in Refs. [1,4,10]. The corresponding Dunkl harmonic oscillator and the annihilation and creation operators take the forms [15]

$$H_\nu = -\frac{1}{2}Y_\nu^2 + \frac{1}{2}x^2 = -\frac{1}{2}\frac{d^2}{dx^2} - \frac{\nu}{x}\frac{d}{dx} + \frac{\nu}{2x^2}(1-R) + \frac{1}{2}x^2, \tag{5}$$

$$A_- = \frac{1}{\sqrt{2}}(Y_\nu + x), \quad A_+ = \frac{1}{\sqrt{2}}(-Y_\nu + x). \tag{6}$$

They satisfy the (anti)commutation relations

$$[A_-, A_+] = 1 + 2\nu R, \quad R^2 = 1, \quad \{A_\pm, R\} = 0, \quad [1, A_\pm] = [1, R] = 0. \tag{7}$$

The generators 1, $A_\pm$, R, and relations (7) give us a realization of the R-deformed Heisenberg algebra [1,10]. In [9,13], the authors show that the R-deformed algebra is intimately related to parabosons, parafermions [13] and to the $osp(1|2)$ $osp(2|2)$ superalgebras.

From now, we assume that $\nu > 0$. The adjoint Y_ν^* of the Dunkl operators Y_ν with domain $\mathcal{S}(\mathbb{R})$ (the space $\mathcal{S}(\mathbb{R})$ being dense in $L^2(\mathbb{R}, |x|^{2\nu}\,dx)$) is $-Y_\nu$ and therefore the operator H_ν is self-adjoint, its spectrum is discrete, and the wave functions corresponding to the well-known eigenvalues

$$\lambda_n = n + \nu + \frac{1}{2}, \quad n = 0, 1, 2, \cdots \tag{8}$$

are given by

$$\psi_n^{(\nu)}(x) = \gamma_n^{-1/2} e^{-x^2/2} H_n^{(\nu)}(x), \tag{9}$$

where

$$\gamma_n = 2^{2n}\Gamma([\tfrac{n}{2}]+1)\Gamma([\tfrac{n+1}{2}]+\nu+\frac{1}{2}), \ n = 0, 1, 2, \cdots.$$

$[x]$ denotes the greatest integer function and $H_n^{(\nu)}(x)$ is the generalized Hermite polynomial introduced by Szegö [15–17] and obtained from Laguerre polynomial $L_n^{(\nu)}(x)$ as follows:

$$\begin{cases} H_{2n}^{(\nu)}(x) = (-1)^n 2^{2n} n! \, L_n^{(\nu-\frac{1}{2})}(x^2), \\ H_{2n+1}^{(\nu)}(x) = (-1)^n 2^{2n+1} n! \, x \, L_n^{(\nu+\frac{1}{2})}(x^2). \end{cases}$$

It is well known that for $\nu > 0$, these polynomials satisfy the orthogonality relations :

$$\int_{\mathbb{R}} H_n^{(\nu)}(x) H_m^{(\nu)}(x) |x|^{2\nu} e^{-x^2}\,dx = \gamma_n \delta_{nm}. \tag{10}$$

We define the generalized Klein operator K as a special case of the fractional Dunkl transform $\mathcal{F}_\nu^\alpha$ corresponding to $\alpha = \frac{2\pi}{r}$. That is,

$$K = \mathcal{F}_\nu^{\frac{2\pi}{r}}. \tag{11}$$

It is well known that the wave functions $\psi_n^{(\nu)}(x)$ form an orthonormal basis of $L^2(\mathbb{R}, |x|^{2\nu}\,dx)$ and are also eigenfunctions of the Fourier–Dunkl transform [11,12,15]. In particular, the generalized Klein operator K acts on the wave functions $\psi_n^{(\nu)}(x)$ as:

$$K\psi_n(x) = \varepsilon_r^n \psi_n^{(\nu)}(x), \quad \varepsilon_r = e^{\frac{2i\pi}{r}}.$$

Let us consider the $\mathbb{Z}_r$-grading structure on the space $L^2(\mathbb{R}, |x|^{2\nu} dx)$ as

$$L^2(\mathbb{R}, |x|^{2\nu} dx) = \bigoplus_{j=0}^{r-1} L_j^2(\mathbb{R}, |x|^{2\nu} dx), \tag{12}$$

where $L_j^2(\mathbb{R}, |x|^{2\nu} dx)$ is a linear subspace of $L^2(\mathbb{R}, |x|^{2\nu} dx)$ generated by the generalized wave functions

$$\{\psi_{nr+j}^{(\nu)}(x) : n = 0, 1, 2, \cdots \}.$$

For $j = 0, 1, \cdots, r-1$, we denote by Π_j, the orthogonal projection from $L^2(\mathbb{R}, |x|^{2\nu} dx)$ onto its subspace $L_j^2(\mathbb{R}, |x|^{2\nu} dx)$. The action of Π_j on $L^2(\mathbb{R}, |x|^{2\nu} dx)$ can be taken to be

$$\Pi_k \psi_{nr+j}^{(\nu)}(x) = \delta_{kj} \psi_{nr+j}^{(\nu)}(x).$$

It is clear that they form a system of resolution of the identity:

$$\Pi_0 + \Pi_1 + \cdots + \Pi_{r-1} = 1, \quad \Pi_i \Pi_j = \delta_{ij} \Pi_i, \quad \Pi_j^* = \Pi_j. \tag{13}$$

Note that the orthogonal projection Π_j is related to the Klein operator K by

$$\Pi_j = \frac{1}{r} \sum_{l=0}^{r-1} \varepsilon_r^{-lj} K^l.$$

3. Fractional Supersymmetric Dunkl Harmonic Oscillator

In this section, we shall present a construction of the fractional supersymmetric quantum mechanics of order r ($r = 2, 3, \ldots$) by using the generalized Klein's operator defined in Equation (11). Following Khare [6,7], a fractional supersymmetric quantum mechanics model of arbitrary order r can be developed by generalizing the fundamental Equations (1) to the forms

$$Q_{\pm}^r = 0, \quad [H, Q_{\pm}] = 0, \quad Q_-^{\dagger} = Q_+,$$
$$Q_-^{r-2} H = Q_-^{r-1} Q_+ + Q_-^{r-2} Q_+ Q_- + \cdots + Q_- Q_+ Q_-^{r-2} + Q_+ Q_-^{r-1}.$$

We introduce the supercharges Q_- and Q_+ as :

$$Q_- = \frac{1}{\sqrt{2}}(Y_\nu + x)(1 - \Pi_0), \qquad Q_+ = \frac{1}{\sqrt{2}}(1 - \Pi_0)(-Y_\nu + x) \tag{14}$$

and the fractional supersymmetric Dunkl harmonic oscillator $\mathcal{H}_\nu$ by

$$\mathcal{H}_\nu = -(r-1)\frac{1}{2}Y_\nu^2 + (r-1)\frac{1}{2}x^2 - \sum_{k=0}^{r-1} \Theta_k \Pi_{r-k-1}, \tag{15}$$

where

$$\Theta_k = \frac{(r-1)(r-2k)}{2} + 2\nu\left[\frac{2r + (-1)^k - 1}{4}\right] R, \quad k = 0, \cdots, r-1, \tag{16}$$

and recall that $[.]$ denotes the greatest integer function. Obviously, the operators $Q_{\pm}$ and $\mathcal{H}_\nu$ with common domain $\mathcal{S}(\mathbb{R})$ are densely defined in the Hilbert space $L^2(\mathbb{R}, |x|^{2\nu} dx)$ and have the Hermitian conjugation relations

$$\mathcal{H}_\nu^* = \mathcal{H}_\nu, \qquad Q_-^* = Q_+. \tag{17}$$

Furthermore, they satisfy the intertwining relations valid for $s = 0, \cdots, r-1$:

$$\Pi_s Q_- = Q_- \Pi_{s+1}, \qquad Q_+ \Pi_s = \Pi_{s+1} Q_+, \qquad \mathcal{H}_v \Pi_s = \Pi_s \mathcal{H}_v. \tag{18}$$

Proposition 1. *The supercharges $Q_\pm$ are nilpotent operators of order r.*

Proof. By making use of the following relations

$$Y_v \Pi_s = \Pi_{s-1} Y_v, \quad x \Pi_s = \Pi_{s+1} x, \tag{19}$$

we can easily show by induction that

$$Q_-^k = \begin{cases} A_-^k \left(1 - \sum_{s=0}^{k-1} \Pi_s\right), & \text{if } 1 \leq k \leq r-1, \\ 0, & \text{if } k = r. \end{cases} \tag{20}$$

Since $Q_+ = Q_-^*$, we also have $Q_+^r = 0$. $\square$

The first main result is

Theorem 1. *The Hermitian operators Q_-, Q_+ and $\mathcal{H}_v$ defined in Equations (14) and (15) satisfy the commutation relations:*

$$(i) \quad Q_\pm^r = 0, \quad [\mathcal{H}_v, Q_\pm] = 0, \quad Q_-^\dagger = Q_+,$$
$$(ii) \quad Q_-^{r-2}\mathcal{H}_v = Q_-^{r-1}Q_+ + Q_-^{r-2}Q_+Q_- + \cdots + Q_-Q_+Q_-^{r-2} + Q_+Q_-^{r-1}.$$

Proof. From the commutation relation (7), we can show by induction that

$$A_+ A_-^k = A_-^k A_+ - \vartheta_k A_-^{k-1}, \quad k \geq 1, \tag{21}$$

where

$$\vartheta_k = \begin{cases} k, & \text{if } k \text{ is even}, \\ k + 2vR, & \text{if } k \text{ is odd}. \end{cases} \tag{22}$$

Combining this with Equation (20), we obtain, for, $k = 1, \cdots, r-2$:

$$\begin{aligned} Q_+ Q_-^{r-1} &= A_-^{r-2}\left(A_- A_+ - \vartheta_{r-1}\right)\Pi_{r-1}, \\ Q_-^{r-1} Q_+ &= A_-^{r-2} A_- A_+ \Pi_{r-2}, \\ Q_-^{r-1-k} Q_+ Q_-^k &= A_-^{r-2}\left(A_- A_+ - \vartheta_k\right)\left(\Pi_{r-2} + \Pi_{r-1}\right). \end{aligned}$$

Additionally, a straightforward computation shows that

$$\sum_{k=1}^{r-1} \vartheta_k = \frac{r(r-1)}{2} + 2v\left[\tfrac{r}{2}\right] R.$$

Thus, we get

$$\sum_{k=0}^{r-1} Q_-^{r-1-k} Q_+ Q_-^k = A_-^{r-2}\left[(r-1)A_- A_+ (\Pi_{r-2} + \Pi_{r-1}) - \left(\sum_{k=1}^{r-2}\vartheta_k\right)\Pi_{r-2} - \left(\sum_{k=1}^{r-1}\vartheta_k\right)\Pi_{r-1}\right]$$
$$= Q_-^{r-2}\left[(r-1)A_- A_+ - \Theta_1\Pi_{r-2} - \Theta_0\Pi_{r-1}\right]. \tag{23}$$

From Equation (13), we easily see that

$$(\Pi_{r-2} + \Pi_{r-1}) \sum_{k=2}^{r-1} \Theta_k \Pi_{r-k-1} = 0,$$

and combining with Equation (23), we get

$$\sum_{k=0}^{r-1} Q_-^{r-1-k} Q_+ Q_-^k = Q_-^{r-2} \mathcal{H}_\nu.$$

It remains to prove that $[\mathcal{H}_\nu, Q_-] = [\mathcal{H}_\nu, Q_+] = 0$. Observe that for $k = 0, \cdots, r-1$, we have

$$r - 1 = \left[\frac{2r + (-1)^k - 1}{4}\right] + \left[\frac{2r + (-1)^{k+1} - 1}{4}\right],$$

and then, for $k = 0, \cdots, r-2$, we have

$$\Theta_k - (r-1)(1 - 2\nu R) = \Theta_{k+1}, \tag{24}$$

which leads to

$$\begin{aligned} Q_- \mathcal{H}_\nu &= \left\{(r-1)A_-A_+ + 1 - 2\nu R - \sum_{k=0}^{r-2} \Theta_k \Pi_{r-k-2}\right\} A_-(1 - \Pi_0) \\ &= \left\{(r-1)A_-A_+ - \sum_{k=0}^{r-2} \Theta_{k+1} \Pi_{r-k-2}\right\}(1 - \Pi_{r-1}) A_- \\ &= \mathcal{H}_\nu Q_-. \end{aligned}$$

Finally, we have obtained $[\mathcal{H}_\nu, Q_-] = 0$, and since the operator $\mathcal{H}_\nu$ is self-adjoint and $Q_+ = Q_-^*$, we conclude that $[\mathcal{H}_\nu, Q_+] = 0$. $\square$

Proposition 2. *For even integer r, the fractional supersymmetric Dunkl harmonic oscillator $\mathcal{H}_\nu$ has $r/2$-fold degenerate spectrum and acts on the wave functions $\psi_n^{(\nu)}(x)$ as:*

$$\mathcal{H}_\nu \psi_{nr+s}^{(\nu)}(x) = \lambda_{nr} \psi_{nr+s}^{(\nu)}(x), \quad s = 0, 1, \ldots r-1, \quad n = 0, 1, 2, \ldots$$

where

$$\lambda_{nr} = (r-1)\left(nr + \nu + \frac{r+1}{2}\right) + (-1)^s \nu r, \quad s = 0, \ldots, r-1.$$

Proof. From ([15] [formulas (3.7.1) and (3.7.2)]), the creation and annihilation operators A_+ and A_- act on the wave functions ψ_{nr+s}^ν as:

$$A_- \psi_{nr+s}^\nu = \sqrt{nr + s + \nu(1 - (-1)^s)}\, \psi_{nr+s-1}^\nu,$$

$$A_+ \psi_{nr+s}^\nu = \sqrt{nr + s + 1 + \nu(1 - (-1)^{s+1})}\, \psi_{nr+s+1}^\nu.$$

Then, the supercharges Q_- and Q_+ take the value

$$Q_- \psi_{nr+s}^\nu = \sqrt{\left(nr + s + \nu(1 - (-1)^s)\right)/2}\, \psi_{nr+s-1}^\nu, \quad s = 1, \cdots, r-1, \tag{25}$$

$$Q_+ \psi_{nr+s}^\nu = \sqrt{\left(nr + s + 1 + \nu(1 - (-1)^{s+1})\right)/2}\, \psi_{nr+s+1}^\nu, \quad s = 0, \cdots, r-2, \tag{26}$$

$$Q_- \psi_{nr}^\nu = 0, \quad Q_+ \psi_{(n+1)r-1}^\nu = 0. \tag{27}$$

A straightforward computation shows that

$$\mathcal{H}_v \psi^v_{nr+s} = \lambda_{nr} \psi^v_{nr+s}, \quad s = 0, \cdots, r-1,$$

where $\lambda_{nr} = (r-1)(nr + v + \frac{r+1}{2}) + (-1)^s vr$. $\quad\square$

4. Supersymmetric Generalized Hermite Polynomials

4.1. Associated Generalized Hermite Polynomials

Starting form the following recurrence relations for the generalized Hermite polynomials $\{H_n^{(v)}(x)\}$,

$$\begin{aligned}
H_{n+1}^{(v)}(x) &= 2xH_n^{(v)}(x) - 2\left(n + v(1 - (-1)^n)\right)H_{n-1}^{(v)}(x) \\
H_0^{(v)}(x) &= 1, \quad H_1^{(v)}(x) = 2x,
\end{aligned} \tag{28}$$

given in [15–17], one defines, for each real number c, the system of polynomials $H_n^{(v)}(x,c)$ by the recurrence relation:

$$H_{n+1}^{(v)}(x,c) = 2xH_n^{(v)}(x,c) - 2\left(n + c + v(1 - (-1)^n)\right)H_{n-1}^{(v)}(x,c), \tag{29}$$

with initial conditions

$$H_0^{(v)}(x,c) = 1, \quad H_1^{(v)}(x,c) = 2x. \tag{30}$$

Now, assume that

$$c > 0, \quad c + 2v > -1. \tag{31}$$

By Favard's theorem [16], it follows that the family of polynomials $\{H_n^{(v)}(x,c)\}$ satisfying the recurrence relation (29) and the initial condition (30), is orthogonal with respect to some positive measure on the real line. We shall refer to the polynomials $\{H_n^{(v)}(x,c)\}$ as the associated generalized Hermite polynomials. As shown in ([18] Theorem 5.6.1)(see also [19–21]), there are two different systems of associated Laguerre polynomials denoted by $L_n^{(v)}(x,c)$ and $\mathcal{L}_n^{(v)}(x,c)$. They satisfy the recurrence relations:

$$(2n + 2c + v + 1 - x)L_n^{(v)}(x,c) = (n + c + 1)L_{n+1}^{(v)}(x,c) + (n + c + v)L_{n-1}^{(v)}(x,c), \tag{32}$$

$$L_0^{(v)}(x,c) = 1, \quad L_1^{(v)}(x,c) = \frac{2c + v + 1 - x}{c + 1} \tag{33}$$

and

$$(2n + 2c + v + 1 - x)\mathcal{L}_n^{(v)}(x,c) = (n + c + 1)\mathcal{L}_{n+1}^{(v)}(x,c) + (n + c + v)\mathcal{L}_{n-1}^{(v)}(x,c), \tag{34}$$

$$\mathcal{L}_0^{(v)}(x,c) = 1, \quad \mathcal{L}_1^{(v)}(x,c) = \frac{c + v + 1 - x}{c + 1}. \tag{35}$$

Recall the Tricomi function $\Psi(a,c;x)$ given by

$$\Psi(a,c;x) = \frac{1}{\Gamma(a)} \int_0^\infty e^{-xt} t^{a-1}(1+t)^{c-a-1}\, dt, \quad \Re(a), \Re(x) > 0.$$

By [18], the polynomials $L_n^{(v)}(x,c)$ and $\mathcal{L}_n^{(v)}(x,c)$ satisfy the orthogonality relations

$$\int_0^\infty L_n^{(v)}(x,c)L_m^{(v)}(x,c)\,x^v e^{-x}\frac{|\Psi(c,1-v;xe^{-i\pi})|^{-2}}{\Gamma(c+1)\Gamma(v+c+1)}\,dx \;=\; \frac{(v+c+1)_n}{(c+1)_n}\delta_{nm}, \tag{36}$$

$$\int_0^\infty \mathcal{L}_n^{(v)}(x,c)\mathcal{L}_m^{(v)}(x,c)\,x^v e^{-x}\frac{|\Psi(c,-v;xe^{-i\pi})|^{-2}}{\Gamma(c+1)\Gamma(v+c+1)}\,dx \;=\; \frac{(v+c+1)_n}{(c+1)_n}\delta_{nm}, \tag{37}$$

when one of the following conditions is satisfied:

$$v+c>-1,\quad c\ge 0 \qquad \text{or} \qquad v+c\ge -1,\quad c\ge -1.$$

The monic polynomial version of $H_n^v(x,c)$ is given by

$$\mathcal{H}_n^{(v)}(x,c)=2^{-n}H_n^{(v)}(x,c),\quad n=0,1,\cdots,$$

and satisfies

$$\mathcal{H}_{n+1}^{(v)}(x,c)=x\mathcal{H}_n^{(v)}(x,c)-\frac{1}{2}\left(n+c+v(1-(-1)^n)\right)\mathcal{H}_{n-1}^{(v)}(x,c),$$
$$\mathcal{H}_{-1}^{(v)}(x,c)=0,\quad \mathcal{H}_0^{(v)}(x,c)=1. \tag{38}$$

It is easy to see that the polynomial $(-1)^n\mathcal{H}_n^{(v)}(-x,c)$ also satisfies (38). Thus,

$$\mathcal{H}_n^{(v)}(-x,c)=(-1)^n\mathcal{H}_n^{(v)}(x,c).$$

Thus, by induction, we write them in the form

$$\mathcal{H}_{2n}^{(v)}(x,c)=S_n(x^2)\quad\text{and}\quad \mathcal{H}_{2n+1}^{(v)}(x,c)=xQ_n(x^2), \tag{39}$$

where $S_n(x)$, $Q_n(x)$ are monic polynomials of degree n.

Theorem 2. *Let $c>0$ and $v>-c/2$. The associated generalized Hermite polynomials $H_n^{(v)}(x,c)$, defined in (29), have the explicit form:*

$$H_{2n}^{(v)}(x,c)=(-1)^n 2^{2n}(1+c/2)_n L_n^{(v-1/2)}(x^2,c/2),$$
$$H_{2n+1}^{(v)}(x,c)=(-1)^n 2^{2n+1}(1+c/2)_n x L_n^{(v+1/2)}(x^2,c/2),$$

and the orthogonality relations

$$\int_{\mathbb{R}} H_n^{(v)}(x,c)H_m^{(v)}(x,c)\,|x|^{2v}e^{-x^2}\frac{|\Psi(c/2,1/2-v;x^2e^{-i\pi})|^{-2}}{\Gamma(1+c/2)\Gamma(v+c/2+1/2)}=\zeta_n\,\delta_{nm}, \tag{40}$$

where

$$\zeta_n=\begin{cases}2^{4k}(1+c/2)_k(v+c/2+1/2)_k, & \text{if}\quad n=2k,\\ 2^{4k+2}(1+c/2)_k(v+c/2+3/2)_k, & \text{if}\quad n=2k+1.\end{cases}$$

Proof. It is directly verified that the polynomials $S_n(x)$, $Q_n(x)$ given in (39) are orthogonal as they satisfy the recurrence relations

$$
\begin{aligned}
S_{n+1}(x) &= (x - (2n + c + v + 1/2))S_n(x) - (n + c/2) \\
&\quad \times (n + c/2 - 1/2 + v)S_{n-1}(x), \\
S_{-1}(x) &= 0, \quad S_0(x) = 1,
\end{aligned}
$$

and

$$
\begin{aligned}
Q_{n+1}(x) &= (x - (2n + c + 3/2 + v))Q_n(x) - (n + c/2)big) \\
&\quad \times (n + (1 + c)/2 + v)Q_{n-1}(x) \\
Q_{-1}(x) &= 0, \quad Q_0(x) = 1.
\end{aligned}
$$

From Equation (32), we see that the polynomials $S_n(x)$ satisfy the same recurrence relation as $(-1)^n(1 + c/2)_n \mathcal{L}_n^{(v-1/2)}(x, c/2)$, so that

$$
S_n(x) = (-1)^n(1 + c/2)_n \mathcal{L}_n^{(v-1/2)}(x, c/2). \tag{41}
$$

A similar analysis shows that

$$
Q_n(x) = (-1)^n(1 + c/2)_n L_n^{(v+1/2)}(x, c/2). \tag{42}
$$

In view of Equations (41) and (42), the explicit form of the associated generalized Hermite polynomials is given by

$$
H_{2n}^{(v)}(x, c) = (-1)^n 2^{2n}(1 + c/2)_n \mathcal{L}_n^{(v-1/2)}(x^2, c/2), \tag{43}
$$

$$
H_{2n+1}^{(v)}(x, c) = (-1)^n 2^{2n+1}(1 + c/2)_n x L_n^{(v+1/2)}(x^2, c/2). \tag{44}
$$

From Equations (36) and (37), we deduce that the system $\mathcal{H}_n^v(x, c)$ satisfies the orthogonality relations

$$
\int_{\mathbb{R}} H_n^{(v)}(x, c) H_m^{(v)}(x, c) |x|^{2v} e^{-x^2} \frac{|\Psi(c/2, 1/2 - v; x^2 e^{-i\pi})|^{-2}}{\Gamma(1 + c/2)\Gamma(v + c/2 + 1/2)} = \zeta_n \delta_{nm}, \tag{45}
$$

with

$$
\zeta_n = \begin{cases} 2^{4k}(1 + c/2)_k(v + c/2 + 1/2)_k, & \text{if } n = 2k, \\ 2^{4k+2}(1 + c/2)_k(v + c/2 + 3/2)_k, & \text{if } n = 2k + 1. \end{cases}
$$

$\square$

4.2. Supersymmetric Generalized Hermite Polynomials

In the sequel, we assume that r is an even integer and we consider the Hermitian supercharge operator Q, defined on $S(\mathbb{R})$, by

$$
Q = \frac{1}{\sqrt{2}} Y_v(\Pi_{r-1} - \Pi_0) + \frac{x}{\sqrt{2}}(2 - \Pi_0 - \Pi_{r-1}).
$$

From Equation (14), we have

$$
Q = \frac{1}{\sqrt{2}}(Q_- + Q_+),
$$

so it has a self-adjoint extension on $L^2(\mathbb{R}, |x|^{2\nu}dx)$. Furthermore, it acts on the basis ψ_n^ν as

$$Q\psi_{nr+s}^\nu = a_s^{(n)}\psi_{nr+s-1}^\nu + a_{s+1}^{(n)}\psi_{nr+s+1}^\nu, \quad s = 1, \cdots, r-1,$$
$$Q\psi_{nr}^\nu = a_1^{(n)}\psi_{nr+1}^\nu, \quad Q\psi_{(n+1)r-1}^\nu = a_{r-1}^{(n)}\psi_{(n+1)r-1}^\nu, \tag{46}$$

where

$$a_s^{(n)} := \sqrt{(nr+s+\nu(1-(-1)^s))/2}, \qquad s = 1, \cdots, r-1.$$

On the other hand, by (46), we see that the operator Q leaves invariant the finite dimensional subspace of $L^2(\mathbb{R}, |x|^{2\nu}dx)$ generated by ψ_{nr+s}^ν, $s = 0, 1, \cdots, r-1$. Hence, Q can be represented in this basis by the following $r \times r$ tridiagonal Jacobi matrix $A_r^{(n)}$

$$A_r^{(n)} = \begin{pmatrix} 0 & a_1^{(n)} & 0 & & & & \\ a_1^{(n)} & 0 & a_2^{(n)} & 0 & & & \\ 0 & a_2^{(n)} & 0 & a_3^{(n)} & \ddots & & \\ & \ddots & \ddots & \ddots & \ddots & 0 & \\ & & \ddots & a_{r-2}^{(n)} & 0 & a_{r-1}^{(n)} \\ & & & 0 & a_{r-1}^{(n)} & 0 \end{pmatrix}.$$

It is well known that, if the coefficients of the subdiagonal of some Jacobi Matrix are different from zero, then all the eigenvalues of this matrix are real and nondegenerate [16]. We introduce the normalized eigenvectors ϕ_s of the supercharge Q

$$Q\phi_s = x_s\phi_s, \qquad s = 0, \cdots, r-1 \tag{47}$$

that can be expanded in the basis ψ_{nr+k}, $k = 0, 1, \cdots, r-1$, as

$$\phi_s = \sum_{k=0}^{r-1} \sqrt{w_s}p_k(x_s)\psi_{nr+k}, \tag{48}$$

where the coefficients p_k obey the three-term recurrence relation [22]

$$a_k^{(n)}p_{k-1}(x) + a_{k+1}^{(n)}p_{k+1}(x) = xp_k(x),$$
$$p_{-1}(x) = 0, \quad p_0(x_s) = 1,$$

Hence, they become orthogonal polynomials. We denote by $P_k(x)$, the monic orthogonal polynomial related to $p_k(x)$ by

$$P_k(x) = h_k p_k(x), \tag{49}$$

where

$$h_k = a_k^{(n)} \cdots a_1^{(n)} \tag{50}$$

and satisfying

$$xP_k(x) = P_{k+1}(x) + \frac{1}{2}(k+nr+\nu(1-(-1)^k))P_{k-1}(x), \quad k = 0, \cdots, r-1,$$
$$P_{-1}(x) = 0, \quad P_0(x) = 1. \tag{51}$$

From the three terms recurrence relations (51), the polynomials $P_k(x)$ can be identified with the associated generalized Hermite polynomial $\mathcal{H}_k^{(v)}(x,c)$, namely,

$$P_k(x) = \mathcal{H}_k^{(v)}(x,nr).$$

It is well known from the theory of orthogonal polynomials that the eigenvalues of the Jacobi matrix $A_r^{(n)}$ coincide with the roots of the characteristic polynomial $\mathcal{H}_r^{(v)}(x,nr)$ [16,22]. The weights w_s defined in (56) are given by the following formula

$$w_s = \frac{h_r^2}{\mathcal{H}_{r-1}^{(v)}(x_s,nr)(\mathcal{H}_r^{(v)})'(x_s,nr)}, \tag{52}$$

where $(\mathcal{H}_r^{(v)})'(x,nr)$ denotes the derivative of $\mathcal{H}_r^{(v)}(x,nr)$, h_r is defined in Equation (50) and $x_{nr,1} > \cdots > x_{nr,r}$ are the zeros of $\mathcal{H}_r^{(v)}(x,nr)$. For more detail, we refer to [16]. Then, it turns out that

$$\phi_s = \sum_{k=0}^{r-1} u_{ks}^{(n)} \psi_{nr+k}, \tag{53}$$

where

$$u_{ks}^{(n)} = \frac{h_r}{h_k} \frac{\mathcal{H}_k^{(v)}(x_s,nr)}{(\mathcal{H}_{r-1}^{(v)}(x_s,nr)(\mathcal{H}_r^{(v)})'(x_s,nr))^{1/2}}, \quad 0 \le s,k \le r-1. \tag{54}$$

Since both bases $\{\psi_{nr+k}, k=0,\cdots r-1\}$ and $\{\phi_s, s=0,\cdots r-1\}$ are orthonormal and all the coefficients are real, then the matrix $(u_{ks}^{(n)})$ is orthogonal and hence the system $\{\mathcal{H}_k^{(v)}(x)\}$ becomes orthogonal polynomials:

$$\sum_{s=0}^{r-1} w_s \mathcal{H}_k^{(v)}(x_s)\mathcal{H}_{k'}^{(v)}(x_s) = \delta_{kk'}/h_k^2. \tag{55}$$

We call supersymmetric generalized Hermite polynomials the orthogonal polynomials, denoted by $\mathbb{H}_N^{(r,v)}(x)$, extracted form the orthogonal function ϕ_s:

$$\mathbb{H}_N^{(r,v)}(x) = \sum_{k=0}^{r-1} H_k^{(v)}(x_s,nr)H_{nr+k}^{(v)}(x), \quad N=nr+s, \tag{56}$$

and we obtain the following:

Theorem 3. *The supersymmetric generalized Hermite polynomials $\mathbb{H}_N^{(r,v)}(x)$ satisfy the orthogonality relations*

$$\int_{-\infty}^{\infty} \mathbb{H}_N^{(r,v)}(x)\mathbb{H}_{N'}^{(r,v)}(x)|x|^{2v}e^{-x^2}\,dx = \varrho_N\delta_{NN'}, \tag{57}$$

where $\varrho_N = \gamma_{nr}/w_s$ for $s=0,\cdots r-1$ and $N=nr+s$.

Proof. From Equations (10) and (56), we obtain

$$\int_{-\infty}^{\infty} \mathbb{H}_{nr+s}^{(r,v)}(x)\mathbb{H}_{n'r+s'}^{(r,v)}(x)\,|x|^{2v}e^{-x^2}\,dx = \delta_{nn'}\sum_{k=0}^{r-1} H_k^{(v)}(x_s,nr)H_k^{(v)}(x_{s'},nr)\gamma_{nr+k}$$

$$= \delta_{nn'}\gamma_{nr}\sum_{k=0}^{r-1} h_k^2 H_k^{(v)}(x_s,nr)\mathcal{H}_k^{(v)}(x_{s'},nr)$$

and, from ([18] Theorem 2.11.2), we obtain the dual orthogonality relation for $\{\mathcal{H}_k^{(\nu)}(x)\}$:

$$\sum_{k=0}^{r-1} \mathcal{H}_k^{(\nu)}(x_s)\mathcal{H}_k^{(\nu)}(x_{s'}) \frac{\left[\frac{nr+k}{2}\right]!\,\Gamma\left(\frac{nr+k+1}{2}+\nu+\frac{1}{2}\right)}{\left[\frac{nr}{2}\right]!\,\Gamma\left(\frac{nr+1}{2}+\nu+\frac{1}{2}\right)} = \delta_{ss'}/w_s \tag{58}$$

and, finally,

$$\int_{-\infty}^{\infty} \mathbb{H}_{nr+s}^{(r,\nu)}(x)\mathbb{H}_{n'r+s'}^{(r,\nu)}(x)\,|x|^{2\nu}e^{-x^2}\,dx = \delta_{nn'}\delta_{ss'}\gamma_{nr}/w_s.$$

$\square$

Author Contributions: Formal analysis, F.B. and W.J.; Methodology, F.B. and W.J.; Writing—original draft, F.B. and W.J.; Writing—review and editing, F.B. and W.J. All authors have read and agreed to the published version of the manuscript.

Funding: The authors would like to extend their sincere appreciation to the Deanship of Scientific Research at King Saudi University for funding this Research Group No. (RG-1437-020).

Conflicts of Interest: The authors declare no conflict of interest.

References

1. Plyushchay, M.S. Deformed Heisenberg algebra, fractional spin fields and supersymmetry without fermions *Annals Phys.* **1996**, *245*, 339–360.
2. Witten, E. Dynamical breaking of supersymmetry. *Nucl. Phys. B* **1981**, *188*, 513–554 [CrossRef]
3. Daoud, M.; Hassouni, Y.; Kibler, M. The k-fermions as objects interpolating between fermions and bosons. In *Symmetries in Science X*; Daoud, M., Hassouni, Y., Kibler, M., Gruber, B., Ramek, M., Eds.; Plenum Press: New York, NY, USA, 1998; pp. 63–77.
4. Plyushchay, M.S. Supersymmetry without fermions DFTUZ-94–05. *arXiv* **1994**, arXiv:hep-th/9404081.
5. Post, S.; Vinet, L.; Zhedanov, A. Supersymmetric quantum mechanics with reflections. *J. Phys. A Math. Theor.* **2011**, *44*, 435301. [CrossRef]
6. Cooper, F.; Khare, A.; Sukhatme, U. *Supersymmetry in Quantum Mechanics*; World Scientific: Singapore, 2001
7. Khare, A. Parasupersymmetric quantum mechanics of arbitrary order. *J. Phys. A* **1992**, *25*, L749. [CrossRef]
8. Quesne, C.; Vansteenkiste, N. C_λ-extended harmonic oscillator and (para) supersymmetric quantum mechanics. *Phys. Lett. A* **1998**, *240*, 21. [CrossRef]
9. Plyushchay, M. Hidden nonlinear supersymmetries in pure parabosonic systems, *Int. J. Mod. Phys. A* **2000**, *15*, 3679–3698. [CrossRef]
10. Plyushchay, M.S. Deformed Heisenberg algebra with reflection. *Nuclear Phys. B* **1997**, *491*, 619–634. [CrossRef]
11. Ghazouani, S.; Bouzeffour, F. A fractional power for Dunkl transform. *Bull. Math. Anal. Appl.* **2014**, *6*, 1–30.
12. Ghazouani, S.; Bouzeffour, F. Heisenberg uncertainty principle for a fractional power of the Dunkl transform on the real line. *J. Computat. Appl. Math.* **2016**, *294*, 151–176. [CrossRef]
13. Watson, G.N. *A Treatise on the Theory of Bessel Functions*, 2nd ed.; Cambridge Univ. Press: London, UK; New York, NY, USA, 1944.
14. Dunkl, C.F.; Xu, Y. Orthogonal polynomials of several variables. In *Encyclopedia of Mathematics and Its Applications*; Cambridge University Press: Cambridge, UK, 2001; Volume 81.
15. Rosenblum, M. Generalized Hermite polynomials and the Bose-like oscillator calculus. In *Non Self-Adjoint Operators and Related Topics*; Sheva, B., Ed.; Springer Basel AG: Birkhauser Verlag: Basel, Switzerland, 1994.
16. Chihara, T. An Introduction to Orthogonal Polynomials. In *Dover Books on Mathematics*, Reprint ed.; Gordon and Breach: New York, NY, USA, 2011.
17. Szegö, G. *Orthogonal Polynomials*; American Mathematical Society: Providence, RI, USA, 1939; Volume 23.
18. Ismail, M.E.H. Classical and quantum orthogonal polynomials in one variable. In *Encyclopedia of Mathematics and its Applications*; Cambridge University Press: Cambridge, UK. 2005; ISBN 0521782015.
19. Askey, R.; Wimp, J. Associated Laguerre and Hermite polynomials. *Proc. Roy. Soc. Edinburgh Sect. A* **1984**, *96*, 15–37. [CrossRef]

20. Kim, T.; Kim, D.; Jang, L.-C.; Kwon, H. On differential equations associated with squared hermite polynomials. *J. Comput. Anal. Appl.* **2017**, *23*, 1252–1264.
21. Kim, T.; Kim, D.S.; Dolgy, D.V.; Park, J.-W. Sums of finite products of Legendre and Laguerre polynomials. *Adv. Differ. Equ.* **2018**, *2018*, 277. [CrossRef]
22. Atkinson, F.V. *Discrete and Continuous Boundary Problems*; Academic Press: New York, NY, USA; London, UK, 1964.

Communication

Rational Approximation for Solving an Implicitly Given Colebrook Flow Friction Equation

Pavel Praks [1,2,*] and Dejan Brkić [1,2,3,*]

1   European Commission, Joint Research Centre (JRC), 21027 Ispra, Italy
2   IT4Innovations, VŠB—Technical University of Ostrava, 708 00 Ostrava, Czech Republic
3   Research and Development Center "Alfatec", 18000 Niš, Serbia
*   Correspondence: pavel.praks@vsb.cz (P.P.); dejanrgf@tesla.rcub.bg.ac.rs or dejan.brkic@vsb.cz (D.B.)

Received: 21 November 2019; Accepted: 13 December 2019; Published: 20 December 2019 

Abstract: The empirical logarithmic Colebrook equation for hydraulic resistance in pipes implicitly considers the unknown flow friction factor. Its explicit approximations, used to avoid iterative computations, should be accurate but also computationally efficient. We present a rational approximate procedure that completely avoids the use of transcendental functions, such as logarithm or non-integer power, which require execution of the additional number of floating-point operations in computer processor units. Instead of these, we use only rational expressions that are executed directly in the processor unit. The rational approximation was found using a combination of a Padé approximant and artificial intelligence (symbolic regression). Numerical experiments in Matlab using 2 million quasi-Monte Carlo samples indicate that the relative error of this new rational approximation does not exceed 0.866%. Moreover, these numerical experiments show that the novel rational approximation is approximately two times faster than the exact solution given by the Wright omega function.

Keywords: hydraulic resistance; pipe flow friction; Colebrook equation; Colebrook–White experiment; floating-point computations; approximations; Padé polynomials; symbolic regression

---

1. Introduction

The Colebrook equation [1] for turbulent flow friction is implicitly given with respect to the unknown Darcy flow friction λ, as shown in Equation (1):

$$\frac{1}{\sqrt{\lambda}} = -2 \cdot \log_{10}\left(\frac{2.51}{Re} \cdot \frac{1}{\sqrt{\lambda}} + \frac{\varepsilon}{3.71} \right) \tag{1}$$

where:

λ—Darcy flow friction factor (dimensionless)
Re—Reynolds number, $4000 < Re < 10^8$ (dimensionless)
ε—relative roughness of inner pipe surface, $0 < \varepsilon < 0.05$ (dimensionless)

As a PhD student at the Imperial College in London, Colebrook developed his empirical equation based on the data from his joint experiment with his supervisor, Prof. White [2]. They experimented with flow of air through pipes with different roughness of the inner pipe surface. The experiment by Colebrook and White was described in a scientific journal and published in 1937 [2], while the related empirical equation by Colebrook was published in 1939 [1].

Compared with some other experimental findings [3], the Colebrook equation fits the friction factor within a few dozen percent of error [4]. The Colebrook equation over the last 80 years has been seen by the industry as an informal standard for flow friction calculation and has been very well accepted in everyday engineering practice. Based on the Colebrook equation, Moody developed a diagram that was used before the era of computers for graphical determination of turbulent flow friction [5]. Today, such nomograms have been replaced by explicit approximations, which introduce some value of error [6,7], or by iterative methods [8–10].

In a central computer processor (CPU), transcendental functions such as logarithmic, exponential, or functions with non-integer terms require execution of numerous floating-point operations, and therefore they should be avoided whenever possible [11–18]. Praks and Brkić [19] recently developed a one-log call iterative method, which uses only one computationally demanding function, and even then only in the first iteration (for all succeeding iterations, cheap Padé approximants are used [20,21]). Based on that approach, few very accurate and efficient explicit approximations suitable for coding and for engineering practice have been constructed [22]. In addition, the same authors developed few approximations of the Colebrook equation based on the Wright ω-function, which are among the most accurate to date [23–25]. On the other hand, they contain one or two logarithmic functions, depending on the chosen version [23,24] (these procedures are based on the previous efforts by Praks and Brkić for symbolic regression [26] and by Brkić with Lambert W-function [27–29]).

In this communication, we make a step forward and we offer for the first time a procedure for the approximate solution of the Colebrook equation based only on rational functions. The presented novel rational approximation procedure introduces a relative error of no more than 0.866% for $0 < \varepsilon < 0.05$ and $4000 < \mathrm{Re} < 10^8$ (as used in engineering practice). The rational approximation procedure is suitable for computer codes (open-source code in commercial Matlab 2019a is given in this communication, and in addition, it is compatible with freeware GNU Octave, version 5.1.0).

After introductory Section 1, Section 2 of this communication gives a short overview of mathematical methods used for the proposed rational approximation procedure. Section 3 describes the rational approximation procedure in detail (including error analysis), Section 4 provides software code along with the algorithm to be followed for the rational approximation approach, while Section 5 contains concluding remarks.

2. Mathematics Behind the Proposed Approximation

Our rational approximation approach is based on Padé approximants [20,21], symbolic regression [30,31] and although not used directly, it is inspired by the Wright ω-function, a cognate of the Lambert W-function [32]. To avoid detailed explanations about the Lambert W-function [33], here it should be noted that in this context it is used to transform the Colebrook equation from the shape implicitly given in respect to the unknown flow friction factor to the explicit form [23,24,27–29,34,35].

2.1. Padé Approximants

The ratio of two power series with properly chosen coefficients of the numerator and denominator can approximate very accurately various functions in a narrow zone around the chosen expanding point. For the expressions in the numerator and denominator, Padé approximants [20,21] use rational functions of given order instead of serial expansions. So, in other words, the Padé approximants can estimate functions usually in a narrow zone as the quotient of two polynomials, often has better approximation properties compared with its truncated Taylor series. Being a quotient, the Padé approximants are composed of lower-degree polynomials, where the degree of polynomials can be chosen according to needs. We use Matlab 2019a in order to generate the needed Padé approximants as replacements of the logarithmic function in our rational approximation approach. We do not use expressions with non-integer exponents, because according to Clamond [10], in the software interpretation it is evaluated through one exponential and one logarithmic function (for example $B^\kappa = e^{\kappa \cdot \ln(B)}$, where κ is in most cases a non-integer). The computational complexity of an algorithm describes the amount of resources required to run it, for example the execution time. Winning and Coole [13] performed 100 million calculations for each mathematical operation using random inputs, with each repeated five times, and found that the most efficient operation for addition requires 23.4 s.

According to them, relative effort for computation referred to addition-1 as a reference for the following values: logarithm to base, 10–3.37; fractional exponential, 3.32, cubed root, 2.71; natural logarithm, 2.69; cubed, 2.38; square root, 2.29; squared, 2.18; multiplication, 1.55, division, 1.35; subtraction, 1.18.

2.2. Symbolic Regression

Symbolic regression is a machine learning approach for finding approximate functions based on evolutionary or genetic algorithms [36]. To avoid imposing prior assumptions on the model, symbolic regression has the ability to search through the space of mathematical expressions to look for an approximate function that best fits a given dataset [31].

We used Eureqa [30], a symbolic regression engine, to obtain our final model. HeuristicLab [37], a software environment for heuristic and evolutionary algorithms, including symbolic regression, can be used instead.

3. Routine Based on Polynomial-Form Expressions

3.1. Replacement of Logarithmic Function

The Colebrook equation can be accurately approximated using a rational approximation procedure, as shown in Equation (2):

$$\frac{1}{\sqrt{\lambda}} \approx -0.8686 \cdot (\zeta_1 + \zeta_2) \tag{2}$$

where:

$$\zeta_1 = 0.02087 \cdot r - 0.07659 \cdot p(r) - \frac{0.5994}{p(r) + 3.846} - \frac{0.0007232}{r} - 0.00007489 \cdot r^2 + 0.1391$$

$$\zeta_2 = p(r) - 7.93$$

$$p(r) = \frac{r \cdot (r \cdot (11 \cdot r + 27) - 27) - 11}{r \cdot (r \cdot (3 \cdot r + 27) + 27) + 3}$$

$$r = 2777.77 \cdot \left(\frac{2.51 \cdot p_0}{Re} + \frac{\varepsilon}{3.71} \right)$$

$$p_0 = \frac{2600 \cdot Re}{657.7 \cdot Re + 214600 \cdot Re \cdot \varepsilon + 12970000} - 13.58 \cdot \varepsilon + \frac{0.0001165 \cdot Re}{0.00002536 \cdot Re + Re \cdot \varepsilon + 105.5} + 4.227$$

Consequently, Equation (2) contains only rational functions, where:

$\zeta_1 + \zeta_2$ rational approximation of $\ln\left(\frac{2.51}{Re} \cdot \frac{1}{\sqrt{\lambda}} + \frac{\varepsilon}{3.71} \right)$;

ζ_1 a rational function that corrects error caused by Padé approximant $p(r)$;

ζ_2 shifted Padé approximant $p(r)$, where the shift $-7.93 \approx \ln(0.00036)$ where $0.00036 \approx \frac{1}{2777.77}$;

r argument of $p(r)$;

$p(r)$ Padé approximant of $\ln(r)$ of order /2,3/ at the expansion point $r = 1$;

p_0 starting point;

and where: $\frac{-2}{\ln(10)} \approx \frac{-2}{2.302585093} \approx -0.868$, as $\log_{10}(\varsigma) = \frac{\ln(\varsigma)}{\ln(10)}$.

Function ζ_2 approximates the required logarithmic function $\ln\left(\frac{2.51}{Re} \cdot \frac{1}{\sqrt{\lambda}} + \frac{\varepsilon}{3.71} \right)$, while ζ_1 corrects its error, as r is not always close to the expansion point, because of the large variability of input parameters of the Colebrook equation (Equation (1)). The rational function ζ_1 and also the starting point p_0 were found by symbolic regression software Eureqa [31], whereas the shift -7.93 in ζ_2 was found in order to minimize the error of the Padé approximation of $p(r) \approx \ln(r)$ for the Colebrook equation, as $\ln(0.00036 \cdot r) \approx \ln(r) - 7.93$, where $0.00036 \approx \frac{1}{2777.77}$. Variable precision arithmetic (VPA) at 4 decimal digit accuracy is assumed for ζ_1 and for p_0. The Padé approximant $p(r)$ is given in

Horner nested polynomial form generated in Matlab 2019a. It is of order /2,3/, which means that the polynomial in numerator contains a monomial of highest degree 2, while the denominator of degree 3. Any other suitable Padé polynomial of any other degree in any other form that can substitute the natural logarithm around the needed expansion point can be used, but in such cases, the rational approximation procedure should be tested again, because these changes can affect the value of the final relative error and its distribution.

3.2. Error Analysis

Distribution of the relative error for the proposed rational approximation procedure, Equation (2), is given in Figure 1. The maximal relative error goes up to 0.866% for $0 < \varepsilon < 0.05$ and $4000 < \text{Re} < 10^8$ (as used in engineering practice). The highest error using 2 million input pairs found by the Sobol quasi-Monte Carlo method is for Re = 71987 and $\varepsilon = 3.1711 \cdot 10^{-7}$ [38]. The Colebrook equation is empirical and it follows logarithmic law, so for the procedure with only rational functions, this level of error is acceptable [39]. For example, Pimenta et al. [7] classify the approximation of Sonnad and Goudar [40] with relative error of up to 3.17%. With two logarithms and one non-integer power, this method [40] belongs to the group of approximations with higher performance indexes and precision. Brkić [6] estimates the relative error of Sonnad and Goudar [40] to be up to 0.8%, which is similar error compared with the rational approximation approach presented here; the same methodology as in Brkić [6] is used for Figure 1.

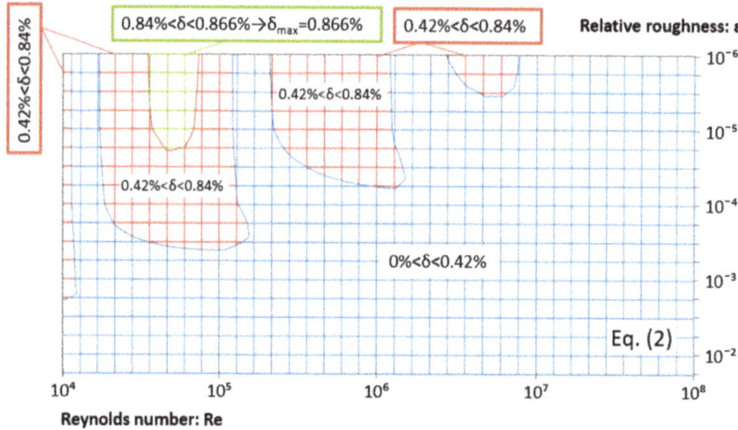

Figure 1. The distribution of the relative error for the proposed rational approximation.

The error in the proposed rational approximation can be possibly reduced by optimizing numerical values of parameters using a genetic algorithms [41] approach with the methodology described by Brkić and Ćojbašić [42]. However, the presented rational approximation approach has too many numerical parameters, meaning such an optimization would be very complex. Further simplifications rather should go in the direction of simplification of p_0, ζ_1 and ζ_2, but keeping the same or increasing accuracy.

3.3. Computational Costs

The efficiency of the proposed procedure is tested using 2 million input pairs found by the Sobol quasi-Monte Carlo method [38]. The tests were performed using Matlab R2019a. The tests revealed that Equation (2) needs 0.56 s to calculate the friction factor λ for 2 million input pairs, or for the Reynolds number *Re* and the roughness of the inner pipe surface ε. On the other hand, the exact solution given by the Wright ω-function [23] implemented by the Matlab library "wrightOmegaq" [43] took 1.1 s for the same 2 million the tested pairs.

Our novel rational approximation of the Colebrook equation given with Equation (2) is approximately two times faster than the exact approach using the Wright ω-function [23], as the speed ratio is 1.1/0.56–1.96. On the other hand, the approach with the Wright ω-function [23] gives the exact solution, which requires two logarithms and one Wright ω-function, while this communication presents a rational approximation.

4. Software Description

The presented rational approximation approach for solving Colebrook's equation for flow friction was thoroughly tested at IT4Innovations, National Supercomputing Center, VŠB-Technical University of Ostrava, Czech Republic.

4.1. Algorithm

The simple algorithm of the rational approximation of the Colebrook equation is presented in Figure 2. The algorithm contains only one branch and is without loops.

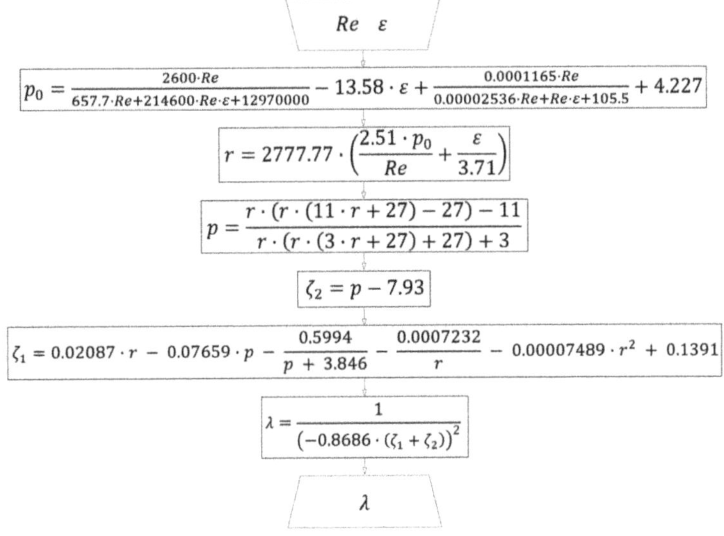

$$p_0 = \frac{2600 \cdot Re}{657.7 \cdot Re + 214600 \cdot Re \cdot \varepsilon + 12970000} - 13.58 \cdot \varepsilon + \frac{0.0001165 \cdot Re}{0.00002536 \cdot Re + Re \cdot \varepsilon + 105.5} + 4.227$$

$$r = 2777.77 \cdot \left(\frac{2.51 \cdot p_0}{Re} + \frac{\varepsilon}{3.71} \right)$$

$$p = \frac{r \cdot (r \cdot (11 \cdot r + 27) - 27) - 11}{r \cdot (r \cdot (3 \cdot r + 27) + 27) + 3}$$

$$\zeta_2 = p - 7.93$$

$$\zeta_1 = 0.02087 \cdot r - 0.07659 \cdot p - \frac{0.5994}{p + 3.846} - \frac{0.0007232}{r} - 0.00007489 \cdot r^2 + 0.1391$$

$$\lambda = \frac{1}{\left(-0.8686 \cdot (\zeta_1 + \zeta_2) \right)^2}$$

Figure 2. Algorithm of the proposed rational approximation of the Colebrook equation.

4.2. Open-Source Software Code

The code is given in Matlab format, which is compatible with the freeware GNU Octave, but it can be easily transposed in any programming language (input parameters: R is the Reynolds number Re, K is the relative roughness of inner pipe surface ε; output parameter; L is the Darcy flow friction factor λ):

$$x = (2600^*R)/(657.7^*R + 214600^*R.^*K + 1.297e{+}7){-}13.58^*K$$
$$+ (1.165e{-}4^*R)/(2.536e{-}5^*R + R.^*K + 105.5) + 4.227$$

$$y_0 = 2.51^*x./R + K./3.71; r = y_0{^*}2777.77$$

$$p_r = (r.^*(r.^*(11^*r + 27) - 27) - 11)/(r.^*(r.^*(3^*r + 27) + 27) + 3)$$

$$k_1 = @(r)\ 0.02087^*r{-}\ 0.07659^*pr{-}\ 0.5994/(pr + 3.846) - 7.232e{-}4./r{-}7.489e{-}5^*r.^2 + 0.1391$$

$$x = -0.8686^*(k_1(r) + pr - 7.93)$$

$$L = 1/x.^2$$

5. Conclusions

We provide a novel rational approximation procedure for solving the logarithmic Colebrook equation for flow friction. Instead of transcendental functions (logarithms, non-integer power) that are used in the classical approach, in this communication, we replace the logarithm with its Padé approximant and a simple rational function, which was found using artificial intelligence (symbolic regression), in order to minimize the error. Although the new rational approximation may seem unintelligible to human eyes, results of 2 million input pairs found by the quasi-Monte Carlo method [38] confirm that the relative error of this new approximation does not exceed 0.866%, which is acceptable for the empirical Colebrook law [44] (trade-off between model complexity and accuracy [45,46]). Consequently, numerical experiments on 2 million of quasi-Monte Carlo pairs indicates that the rational approximation presented here provides for Colebrook's flow friction model a useful combination of Padé approximants and artificial intelligence (symbolic regression).

Author Contributions: P.P. got the idea for the presented rational approximation approach and developed its first version. D.B. put the rational approximation approach into a form suitable for everyday engineering use. D.B. made a draft of this short note and prepared figures. P.P. wrote the software code in consultations with D.B. and tested it. All authors have read and agreed to the published version of the manuscript.

Funding: The work of P.P. has been partially supported by the Technology Agency of the Czech Republic under the project "National Centre for Energy" TN01000007 and "Energy System for Grids" TK02030039. The work of D.B. has been partially supported by the Ministry of Education, Youth, and Sports of the Czech Republic through the National Programme of Sustainability II (NPS II), IT4Innovations Excellence in Science project LQ1602, and by the Ministry of Education, Science, and Technological Development of the Republic of Serbia under the project "Development of new information and communication technologies, based on advanced mathematical methods, with applications in medicine, telecommunications, power systems, protection of national heritage, and education" iii44006.

Acknowledgments: We especially thank Marcelo Masera for his scientific supervision and approval. We acknowledge support from the European Commission, Joint Research Centre (JRC), Directorate C: Energy, Transport, and Climate, Unit C3: Energy Security, Distribution, and Markets.

Notations

The following symbols are used in this Communication:

λ	Darcy, Darcy–Weisbach, Moody, or Colebrook flow friction factor (dimensionless)
Re	Reynolds number, $4000 < Re < 10^8$ (dimensionless)
ε	relative roughness of inner pipe surface, $0 < \varepsilon < 0.05$ (dimensionless)
$\zeta_1 + \zeta_2$	rational approximation of $ln\left(\frac{2.51}{Re} \cdot \frac{1}{\sqrt{\lambda}} + \frac{\varepsilon}{3.71}\right)$
ζ_1	a rational function that corrects error caused by Padé approximant $p(r)$
ζ_2	shifted Padé approximant $p(r)$
r	argument of $p(r)$
$p(r)$	Padé approximant of $\ln(r)$ at the expansion point r
p_0	polynomial starting point
log_{10}	logarithm with base 10
ln	natural logarithm
e	exponential function
ω	Wright ω-function (Wright omega function)

References

1. Colebrook, C.F. Turbulent flow in pipes with particular reference to the transition region between the smooth and rough pipe laws. *J. Inst. Civ. Eng.* **1939**, *11*, 133–156. [CrossRef]
2. Colebrook, C.; White, C. Experiments with fluid friction in roughened pipes. *Proc. R. Soc. Lond. Ser. A Math. Phys. Sci.* **1937**, *161*, 367–381. [CrossRef]

3.   McKeon, B.J.; Zagarola, M.V.; Smits, A.J. A new friction factor relationship for fully developed pipe flow. *J. Fluid Mech.* **2005**, *538*, 429–443. [CrossRef]

4.   Brkić, D.; Praks, P. Unified friction formulation from laminar to fully rough turbulent flow. *Appl. Sci.* **2018**, *8*, 2036. [CrossRef]

5.   Moody, L.F. Friction factors for pipe flow. *Trans. ASME* **1944**, *66*, 671–684.

6.   Brkić, D. Review of explicit approximations to the Colebrook relation for flow friction. *J. Pet. Sci. Eng.* **2011**, *77*, 34–48. [CrossRef]

7.   Pimenta, B.D.; Robaina, A.D.; Peiter, M.X.; Mezzomo, W.; Kirchner, J.H.; Ben, L.H.B. Performance of explicit approximations of the coefficient of head loss for pressurized conduits. *Rev. Bras. Eng. Agríc. Ambient.* **2018**, *22*, 301–307. [CrossRef]

8.   Praks, P.; Brkić, D. Advanced iterative procedures for solving the implicit Colebrook equation for fluid flow friction. *Adv. Civ. Eng.* **2018**, *2018*, 5451034. [CrossRef]

9.   Praks, P.; Brkić, D. Choosing the optimal multi-point iterative method for the Colebrook flow friction equation. *Processes* **2018**, *6*, 130. [CrossRef]

10. Alfaro-Guerra, M.; Guerra-Rojas, R.; Olivares-Gallardo, A. Evaluación experimental de la solución analítica exacta de la ecuación de Colebrook-White (Experimental evaluation of exact analytical solution of the Colebrook-White Equation). *Ing. Investig. Tecnol.* **2019**, *20*, 1–11. (in Spanish). [CrossRef]

11. Clamond, D. Efficient resolution of the Colebrook equation. *Ind. Eng. Chem. Res.* **2009**, *48*, 3665–3671. [CrossRef]

12. Winning, H.K.; Coole, T. Explicit friction factor accuracy and computational efficiency for turbulent flow in pipes. *Flow Turbul. Combust.* **2013**, *90*, 1–27. [CrossRef]

13. Winning, H.K.; Coole, T. Improved method of determining friction factor in pipes. *Int. J. Numer. Methods Heat Fluid Flow* **2015**, *25*, 941–949. [CrossRef]

14. Biberg, D. Fast and accurate approximations for the Colebrook equation. *J. Fluids Eng.* **2017**, *139*, 031401. [CrossRef]

15. Vatankhah, A.R. Approximate analytical solutions for the Colebrook equation. *J. Hydraulic Eng.* **2018**, *144*, 06018007. [CrossRef]

16. Brkić, D.; Praks, P. Discussion of "Approximate analytical solutions for the Colebrook equation". *J. Hydraulic Eng.* **2020**, *146*, 07019011. [CrossRef]

17. Lamri, A.A. Discussion of "Approximate analytical solutions for the Colebrook equation". *J. Hydraulic Eng.* **2020**, *146*, 07019011. [CrossRef]

18. Vatankhah, A.R. Closure to "Approximate analytical solutions for the Colebrook equation". *J. Hydraulic Eng.* **2020**, *146*, 07019013. [CrossRef]

19. Praks, P.; Brkić, D. One-log call iterative solution of the Colebrook equation for flow friction based on Padé polynomials. *Energies* **2018**, *11*, 1825. [CrossRef]

20. Baker, G.A.; Graves-Morris, P. *Padé Approximants*, 2nd ed.; Cambridge University Press: Cambridge, UK, 1996.

21. Copley, L. Padé Approximants, Mathematics for the Physical Sciences. *Walter Gruyter GmbH Co KG* **2014**, *7*, 163–179. [CrossRef]

22. Brkić, D.; Praks, P. Colebrook's flow friction explicit approximations based on fixed-point iterative cycles and symbolic regression. *Computation* **2019**, *7*, 48. [CrossRef]

23. Brkić, D.; Praks, P. Accurate and efficient explicit approximations of the Colebrook flow friction equation based on the Wright ω-function. *Mathematics* **2019**, *7*, 34. [CrossRef]

24. Brkić, D.; Praks, P. Accurate and efficient explicit approximations of the Colebrook flow friction equation based on the Wright ω-function: Reply to Discussion. *Mathematics* **2019**, *7*, 410. [CrossRef]

25. Muzzo, L.E.; Pinho, D.; Lima, L.E.; Ribeiro, L.F. Accuracy/Speed analysis of pipe friction factor correlations. In *International Congress on Engineering and Sustainability in the XXI Century—INCREaSE 2019, Section: Water for Ecosystem and Society, Faro, Portugal, 9–11 October 2019*; Monteiro, J., Silva, A.J., Mortal, A., Aníbal, J., da Silva, M.M., Oliveira, M., Eds.; Springer Nature Switzerland AG 2020: Cham, Switzerland, 2019; pp. 664–679. [CrossRef]

26. Praks, P.; Brkić, D. Symbolic regression-based genetic approximations of the Colebrook equation for flow friction. *Water* **2018**, *10*, 1175. [CrossRef]

27. Brkić, D. W solutions of the CW equation for flow friction. *Appl. Math. Lett.* **2011**, *24*, 1379–1383. [CrossRef]

28. Brkić, D. Lambert W function in hydraulic problems. *Math. Balk.* **2012**, *26*, 285–292. Available online: http://www.math.bas.bg/infres/MathBalk/MB-26/MB-26-285-292.pdf (accessed on 13 November 2019).

29. Brkić, D. Discussion of "Exact analytical solutions of the Colebrook-White equation" by Yozo Mikata and Walter, S. Walczak. *J. Hydraul. Eng.* **2017**, *143*, 07017007. [CrossRef]

30. Dubčáková, R. Eureqa: Software review. *Genet. Program. Evolvable Mach.* **2011**, *12*, 173–178. [CrossRef]

31. Schmidt, M.; Lipson, H. Distilling free-form natural laws from experimental data. *Science* **2009**, *324*, 81–85. [CrossRef]

32. Hayes, B. Why W? *Am. Sci.* **2005**, *93*, 104–108. [CrossRef]

33. Corless, R.M.; Gonnet, G.H.; Hare, D.E.; Jeffrey, D.J.; Knuth, D.E. On the Lambert W function. *Adv. Comput. Math.* **1996**, *5*, 329–359. [CrossRef]

34. Brkić, D. An explicit approximation of Colebrook's equation for fluid flow friction factor. *Pet. Sci. Technol.* **2011**, *29*, 1596–1602. [CrossRef]

35. Brkić, D. Comparison of the Lambert W-function based solutions to the Colebrook equation. *Eng. Comput.* **2012**, *29*, 617–630. [CrossRef]

36. Giustolisi, O.; Savić, D.A. A symbolic data-driven technique based on evolutionary polynomial regression. *J. Hydroinformatics* **2006**, *8*, 207–222. [CrossRef]

37. Wagner, S.; Kronberger, G.; Beham, A.; Kommenda, M.; Scheibenpflug, A.; Pitzer, E.; Vonolfen, S.; Kofler, M.; Winkler, S.; Dorfer, V.; et al. Architecture and design of the HeuristicLab optimization environment. *Top. Intell. Eng. Inform.* **2014**, *6*, 197–261. [CrossRef]

38. Sobol, I.M.; Turchaninov, V.I.; Levitan, Y.L.; Shukhman, B.V. *Quasi-Random Sequence Generators*; Distributed by OECD/NEA Data Bank; Keldysh Institute of Applied Mathematics; Russian Academy of Sciences: Moscow, Russia, 1992; Available online: https://ec.europa.eu/jrc/sites/jrcsh/files/LPTAU51.rar (accessed on 19 November 2019).

39. Eck, B.J.; Mevissen, M. Quadratic approximations for pipe friction. *J. Hydroinformatics* **2015**, *17*, 462–472. [CrossRef]

40. Sonnad, J.R.; Goudar, C.T. Turbulent flow friction factor calculation using a mathematically exact alternative to the Colebrook–White equation. *J. Hydraul. Eng.* **2006**, *132*, 863–867. [CrossRef]

41. Mitrev, R.; Tudjarov, B.; Todorov, T. Cloud-based expert system for synthesis and evolutionary optimization of planar linkages. *Facta Univ. Ser. Mech. Eng.* **2018**, *16*, 139–155. [CrossRef]

42. Brkić, D.; Ćojbašić, Ž. Evolutionary optimization of Colebrook's turbulent flow friction approximations. *Fluids* **2017**, *2*, 15. [CrossRef]

43. Horchler, A.D. Complex Double-Precision Evaluation of the Wright Omega Function. 2017. Available online: https://github.com/horchler/wrightOmegaq (accessed on 10 December 2019).

44. Niazkar, M. Revisiting the estimation of Colebrook friction factor: A comparison between artificial intelligence models and CW based explicit equations. *KSCE J. Civ. Eng.* **2019**, *23*, 4311–4326. [CrossRef]

45. Davidson, J.W.; Savić, D.; Walters, G.A. Method for the identification of explicit polynomial formulae for the friction in turbulent pipe flow. *J. Hydroinformatics* **1999**, *1*, 115–126. [CrossRef]

46. Giustolisi, O.; Berardi, L.; Walski, T.M. Some explicit formulations of Colebrook–White friction factor considering accuracy vs. computational speed. *J. Hydroinformatics* **2010**, *13*, 401–418. [CrossRef]

Article

Iterating the Sum of Möbius Divisor Function and Euler Totient Function

Daeyeoul Kim [1], Umit Sarp [2,*] and Sebahattin Ikikardes [2]

[1] Department of Mathematics and Institute of Pure and Applied Mathematics, Jeonbuk National University, 567 Baekje-daero, deokjin-gu, Jeonju-si, Jeollabuk-do 54896, Korea; kdaeyeoul@jbnu.ac.kr

[2] Faculty of Arts and Science Department of Mathematics, Balikesir University, 10145 Balikesir, Turkey; skardes@balikesir.edu.tr

* Correspondence: umitsarp@ymail.com

Received: 15 October 2019; Accepted: 6 November 2019; Published: 10 November 2019

Abstract: In this paper, according to some numerical computational evidence, we investigate and prove certain identities and properties on the absolute Möbius divisor functions and Euler totient function when they are iterated. Subsequently, the relationship between the absolute Möbius divisor function with Fermat primes has been researched and some results have been obtained.

Keywords: Möbius function; divisor functions; Euler totient function

MSC: 11M36; 11F11; 11F30

---

1. Introduction and Motivation

Divisor functions, Euler φ-function, and Möbius μ-function are widely studied in the field of elementary number theory. The absolute Möbius divisor function is defined by

$$U(n) := |\sum_{d|n} d\mu(d)|.$$

Here, n is a positive integer and μ is the Möbius function. It is well known ([1], p. 23) that

$$\varphi(n) = \sum_{d|n} \mu(d)\frac{n}{d},$$

where φ denotes the Euler φ-function (totient function). If n is a square-free integer, then $U(n) = \varphi(n)$. The first twenty values of $U(n)$ and $\varphi(n)$ are given in Table 1.

Table 1. Values of $U(n)$ and $\varphi(n)$ $(1 \leq n \leq 20)$.

n	1	2	3	4	5	6	7	8	9	10	11	12	13	14	15	16	17	18	19	20
$U(n)$	1	1	2	1	4	2	6	1	2	4	10	2	12	6	8	1	16	2	18	4
$\varphi(n)$	1	1	2	2	4	2	6	4	6	4	10	4	12	6	8	8	16	6	18	8

Let $U_0(n) := n$, $U(n) := U_1(n)$ and $U_m(n) := U_{m-1}(U(n))$, where $m \geq 1$.

Next, to study the iteration properties of $U_m(n)$ (resp., $\varphi_{m'}(n)$), we say the order (resp., class) of n, m-gonal (resp., m'-gonal) absolute Möbius (resp., totient) shape numbers, and shape polygons derived from the sum of absolute Möbius divisor (resp., Euler totient) function are as follows.

Definition 1. *(Order Notion) To study when the positive integer $U_m(n)$ is terminated at one, we consider a notation as follows. The order of a positive integer $n > 1$ denoted $Ord_2(n) = m$, which is the least positive integer m when $U_m(n) = 1$ and $U_{m-1}(n) \neq 1$. The positive integers of order 2 are usually called involutions. Naturally, we define $Ord_2(1) = 0$. The first 20 values of $Ord_2(n)$ and $C(n) + 1$ are given by Table 2. See [2].*

Table 2. Values of $Ord_2(n)$ and $C(n) + 1$ ($1 \leq n \leq 20$).

n	1	2	3	4	5	6	7	8	9	10	11	12	13	14	15	16	17	18	19	20
$Ord_2(n)$	0	1	2	1	2	2	3	1	2	2	3	2	3	3	2	1	2	2	3	2
$C(n) + 1$	0	1	2	2	3	2	3	3	3	3	4	3	4	3	4	4	5	3	4	4

Remark 1. *Define $\varphi_0(n) = n$, $\varphi_1(n) = \varphi(n)$ and $\varphi_k(n) = \varphi(\varphi_{k-1}(n))$ for all $k \geq 2$. Shapiro [2] defines the class number $C(n)$ of n by that integer C such that $\varphi_C(n) = 2$. Some values of $Ord_2(n)$ are equal to them of $C(n) + 1$. Shapiro [2] defined $C(1) + 1 = C(2) + 1 = 1$. Here, we define $C(1) + 1 = 0$ and $C(2) + 1 = 1$. A similar notation of $Ord(n)$ is in [3].*

Definition 2. *(Absolute Möbius m-gonal shape number and totient m′-gonal shape number) If $Ord_2(n) = m - 2$ (resp., $C(n) + 1 = m'$, we consider the set $\{(i, U_i(n)) | i = 0, ..., m - 2\}$ (resp., $\{(i, \varphi_i(n)) | i = 0, ..., m' - 2\}$ and add $(0,1)$. We then put $V_n = \{(i, U_i(n)) | i = 0, ..., m - 2\} \cup \{(0,1)\}$ (resp., $R_n = \{(i, \varphi_i(n)) | i = 0, ..., m' - 2\} \cup \{(0,1)\}$). Then we find a m-gon (resp., m′-gon) derived from V_n (resp., R_n). Here, we call n an absolute Möbius m-gonal shape number (resp., totient m′-gonal shape number derived from U and V_n (resp., φ and R_n) except $n = 1$.*

Definition 3. *(Convexity and Area) We use same notations, convex, non-convex, and area in [3]. We say that n is an absolute Möbius m-gonal convex (resp., non-convex) shape number with respect to the absolute Möbius divisor function U if $\{(i, U_i(n)) | i = 0, ..., m - 2\} \cup \{(0,1)\}$ is convex (resp., non-convex). Let $A(n)$ denote the area of the absolute Möbius m-gon derived from the absolute Möbius m-gonal shape number. Similarly, we define the totient m′-gonal convex (resp., non-convex) shape number and $B(n)$ denote the area of the totient m′-gon.*

Example 1. *If $n = 2$ then we obtain the set of points $V_2 = R_2 = \{(0,2), (1,1), (0,1)\}$. Thus, 2 is an absolute Möbius 3-gonal convex number with $A(2) = \frac{1}{2}$. See Figure 1. See Figures 2–4 for absolute Möbius n-gonal shape numbers and totient n-gonal shape numbers with $n = 2, 3, 4, 5$. The first 19 values of $A(n)$ and $B(n)$ are given by Table 3.*

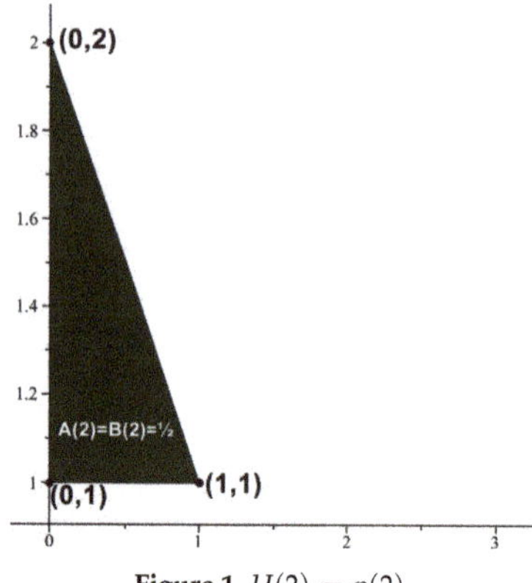

Figure 1. $U(2) = \varphi(2)$.

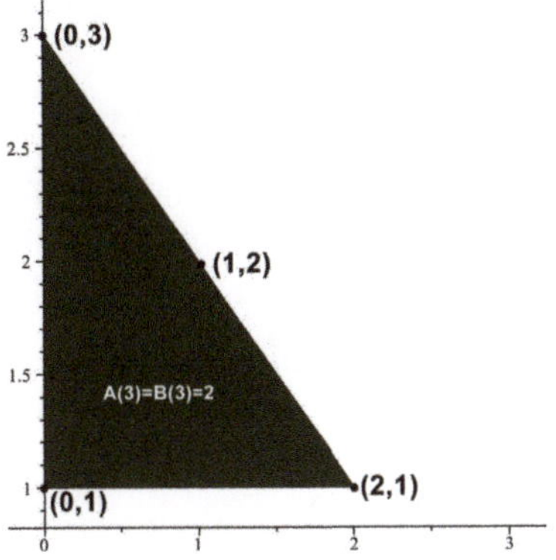

Figure 2. $U(3) = \varphi(3)$.

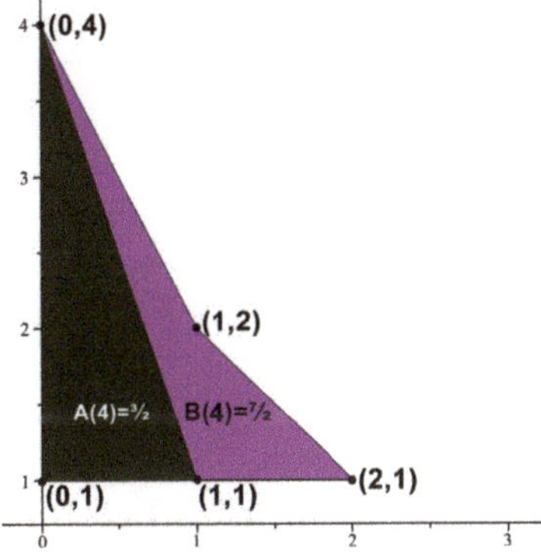

Figure 3. $U(4)$ and $\varphi(4)$.

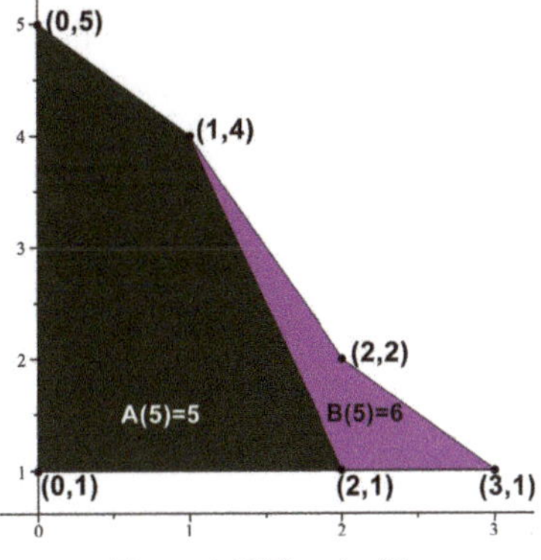

Figure 4. $U(5)$ and $\varphi(5)$.

Table 3. Values of $A(n)$ and $B(n)$ ($2 \leq n \leq 20$).

n	2	3	4	5	6	7	8	9	10	11	12	13	14	15	16	17	18	19	20
$A(n)$	$\frac{1}{2}$	2	$\frac{3}{2}$	5	$\frac{7}{2}$	9	$\frac{7}{2}$	5	$\frac{15}{2}$	17	$\frac{13}{2}$	18	$\frac{25}{2}$	14	$\frac{15}{2}$	23	$\frac{19}{2}$	27	$\frac{25}{2}$
$B(n)$	$\frac{1}{2}$	2	$\frac{5}{2}$	6	$\frac{7}{2}$	9	$\frac{15}{2}$	10	$\frac{17}{2}$	18	$\frac{19}{2}$	21	$\frac{25}{2}$	18	$\frac{37}{2}$	34	$\frac{29}{2}$	32	$\frac{41}{2}$

Kim and Bayad [3] considered the iteration of the odd divisor function S, polygon shape, convex, order, etc.

In this article, we considered the iteration of the absolute Möbius divisor function and Euler totient function and polygon types.

Now we state the main result of this article. To do this, let us examine the following theorem. For the proof of this theorem, the definitions and lemmas in the other chapters of this study have been utilized.

Theorem 1. *(Main Theorem) Let $p_1, \ldots, p_u$ be Fermat primes with $p_1 < p_2 \ldots < p_u$,*

$$F_0 := \{p_1, \ldots, p_u\},$$

$$F_1 := \left\{ \prod_{i=1}^{t} p_i \mid p_i \in F_0, \ 1 \leq t \leq 5 \right\},$$

$$F_2 := \left\{ \prod_{j=1}^{r} p_{i_j} \mid p_{i_j} \in F_0, \ p_1 \leq p_{i_1} < p_{i_2} \ldots < p_{i_r} \leq p_u, \ r \leq u \right\} - (F_0 \cup F_1).$$

If $Ord_2(m) = 1$ or 2 then a positive integer $m > 1$ is

$$\begin{cases} \text{an absolute Möbius 3-gonal (triangular) shape number,} & \text{if } m = 2^k \text{ or } m \in F_1 \\ \text{an absolute Möbius 4-gonal convex shape number,} & \text{if } m \in F_0 - \{3\} \text{ or } m \in F_2 \\ \text{an absolute Möbius 4-gonal non-convex shape number,} & \text{otherwise.} \end{cases}$$

Remark 2. *Shapiro [2] computed positive integer m when $C(m) + 1 = 2$. That is, $m = 3, 4, 6$. Let $C(m) + 1 = 1$ or 2. Then*

(1) If $m = 2, 3$ then m are totient 3-gonal (triangular) numbers.
(2) If $m = 4, 5$ then m are totient 4-gonal non-convex numbers.

2. Some Properties of $U(n)$ and $\varphi(n)$

It is well known [1,4–14] that Euler φ-function have several interesting formula. For example, if $(x, y) = 1$ with two positive integers x and y, then $\varphi(xy) = \varphi(x)\varphi(y)$. On the other hand, if x is a multiple of y, then $\varphi(xy) = y\varphi(x)$ [2]. In this section, we will consider the arithmetic functions $U(n)$ and $\varphi(n)$.

Lemma 1. *Let $n = p_1^{e_1} p_2^{e_2} \cdots p_r^{e_r}$ be a factorization of n, where p_r be distinct prime integers and e_r be positive integers. Then,*

$$U(n) = \prod_{i=1}^{r} (p_i - 1).$$

Proof. If $n = p_1^{e_1} p_2^{e_2} \cdots p_r^{e_r}$ is an arbitrary integer, then we easily check

$$U(n) = \left| \sum_{d \mid n} \mu(d) d \right|$$
$$= \left| 1 - p_1 - p_2 - \ldots - p_r + p_1 p_2 + \ldots + (-1)^n p_1 p_2 \ldots p_r \right|$$
$$= (p_1 - 1)(p_2 - 1) \ldots (p_r - 1).$$

This is completed the proof of Lemma 1. □

Corollary 1. *If p is a prime integer and α is a positive integer, then $U(p) = p - 1$ and $U(p^{\alpha}) = U(p)$. In particular, $U(2^{\alpha}) = 1$.*

Proof. It is trivial by Lemma 1. $\square$

Corollary 2. *Let $n > 1$ be a positive integer and let $Ord_2(n) = m$. Then,*

$$U_0(n) > U_1(n) > U_2(n) > \cdots > U_m(n). \tag{1}$$

Proof. It is trivial by Lemma 1. $\square$

Remark 3. *We compare $U(n)$ with $\varphi(n)$ as follow on Table 4.*

Table 4. $U(n)$ and $\varphi(n)$.

	$U(n)$	$\varphi(n)$
$n = p_1^{e_1} \cdots p_r^{e_r}$	$(p_1 - 1) \cdots (p_r - 1)$	$(p_1^{e_1} - p_1^{e_1-1}) \cdots (p_r^{e_r} - p_r^{e_r-1})$
$n = 2^k$	1	2^{k-1}
sequences	$U_0(n) > U_1(n) > U_2(n) > \cdots$	$\varphi_0(n) > \varphi_1(n) > \varphi_2(n) > \cdots$

Lemma 2. *The function U is multiplicative function. That is, $U(mn) = U(m)U(n)$ with $(m, n) = 1$. Furthermore, if m is a multiple of n, then $U(mn) = U(m)$.*

Proof. Let $m = p_1^{e_1} p_2^{e_2} ... p_i^{e_i}$ and $n = q_1^{f_1} q_2^{f_2} ... q_s^{f_s}$ be positive integers. Then $p_1^{e_1}, p_2^{e_2}, ..., p_i^{e_i}$ and $q_1^{f_1}, q_2^{f_2}, ..., q_s^{f_s}$ are distinct primes. If $(m, n) = 1$ and also $p \mid m$, $p n$, $q \mid n$, and $q m$ by Lemma 1, we note that

$$U(mn) = \prod_{t_k \mid mn} (t_k - 1)$$
$$= \prod_{p_i \mid m} (p_i - 1) \prod_{q_s \mid n} (q_s - 1)$$
$$= U(m) U(n).$$

Let m be a multiple of n. If $p_i \mid n$ then $p_i \mid m$. Thus, by Lemma 1, $U(mn) = U(m)$. This is completed the proof of Lemma 2. $\square$

Remark 4. *Two functions $U(n)$ and $\varphi(n)$ have similar results as follows on Table 5. Here, $n \mid m$ means that m is a multiple of n.*

Table 5. $U(n)$ and $\varphi(n)$.

	$(m, n) = 1$	$n \mid m$
$U(mn)$	$U(m)U(n)$	$U(m)$
$\varphi(mn)$	$\varphi(m)\varphi(n)$	$n\varphi(m)$

Theorem 2. *For all $n \in \mathbb{N} - \{1\}$, there exists $m \in \mathbb{N}$ satisfying $U_m(n) = 1$.*

Proof. Let $n = p_1^{e_1} p_2^{e_2} ... p_r^{e_r}$, where $p_1, ..., p_r$ be distinct prime integers with $p_1 < p_2 < ... < p_r$.

We note that $U(n) = \prod_{i=1}^{r} (p_i - 1)$ by Lemma 1.

If $r = 1$ and $p_1 = 2$, then $U(n) = 1$ by Corollary 1.

If p_i is an odd positive prime integer, then $U(p_i^{e_i}) = p_i - 1$ by Corollary 1.

We note that $p_i - 1$ is an even integer. Then there exist distinct prime integers $q_{i_1}, ..., q_{i_s}$ satisfying

$$p_i - 1 = 2^{l_i} q_{i_1}^{f_{i_1}} \cdots q_{i_s}^{f_{i_s}},$$

where $f_{i_s} \geq 1$, $l_i \geq 1$ and $q_{i_1} < \cdots < q_{i_s}$. It is well known that $q_{i_s} \leq \frac{p_i - 1}{2} \leq \frac{p_r - 1}{2}$.

By Lemma 2, we get

$$U_2(p_i^{e_i}) = U(p_i - 1) = U(q_{i_1}) \cdots U(q_{i_s}). \tag{2}$$

By using the same method in (2) for $1 \leq i_j \leq s$, we get

$$
\begin{aligned}
U_2(n) &= U(\prod_{i=1}^{r} (p_i - 1)) \\
&= U\left(2^{l_1 + l_2 + ... + l_r} \prod_{i=1}^{r} q_{i_1}^{e_{i_1}} ... q_{i_u}^{e_{i_u}} \right) \\
&= U\left(\prod_{i=1}^{r} q_{i_1}^{e_{i_1}} ... q_{i_u}^{e_{i_u}} \right) \\
&= \prod_{i=1}^{r} (q_{i_1} - 1) \cdots (q_{i_u} - 1) \\
&= q_{j_1}^{(2)} \cdots q_{j_k}^{(2)}
\end{aligned}
$$

with $q_{j_1}^{(2)} < q_{j_2}^{(2)} < ... < q_{j_k}^{(2)}$. It is easily checked that $q_{j_k}^{(2)} \leq \max\{\frac{q_{1_u} - 1}{2}, ..., \frac{q_{r_u} - 1}{2}\}$.

Using this technique, we can find l satisfying

$$U_{l-1}(n) = \left(q_{j_1}^{(l-1)} - 1 \right) \cdots \left(q_{j_u}^{(l-1)} - 1 \right) = 2^h \prod_{u=1}^{s'} q_{j_u}^{(l)}$$

with $q_{j_u}^{(l)} < 100$.

By Appendix A (Values of $U(n)$ ($1 \leq n \leq 100$)), we easily find a positive integer v that $U_v(n') = 1$ for $1 \leq n' \leq 100$. Thus, we get $U_v(U_{l-1}(n)) = 1$. Therefore, we can find $m = v + l - 1 \in \mathbb{N}$ satisfying $U_m(n) = 1$. $\square$

Corollary 3. *For all $n \in \mathbb{N} - \{1\}$, there exists $m \in \mathbb{N}$ satisfying $Ord(n) = m$.*

Proof. It is trivial by Theorem 2. $\square$

Remark 5. *Kim and Bayad [3] considered iterated functions of odd divisor functions $S_m(n)$ and order of n. For order of divisor functions, we do not know $Ord(n) = \infty$ or not. But, functions $U_m(n)$ (resp., $\varphi_l(n)$), we know $Ord(n) < \infty$ by Corollary 3 (resp., [15]).*

Theorem 3. *Let $n > 1$ be a positive integer. Then $Ord_2(n) = 1$ if and only if $n = 2^k$ for some $k \in \mathbb{N}$.*

Proof. ($\Leftarrow$) Let $n = 2^k$. It is easy to see that $U(n) = U_1(n) = 1$.

($\Rightarrow$) Let $n = p_1^{e_1} p_2^{e_2} ... p_r^{e_r}$ be a factorization of n, and all p_r are distinct prime integers. If $Ord_2(n) = 1$, then by using Lemma 1 we can note that,

$$1 = (p_1 - 1)(p_2 - 1) ... (p_r - 1). \tag{3}$$

According to all p_r are distinct prime integers, then it is easy to see that there is only exist p_1 and that is $p_1 = 2$. Hereby $n = 2^k$ for some $k \in \mathbb{N}$.

This is completed the proof of Theorem 3. $\square$

Remark 6. *If $k > 0$ then 2^k is an absolute Möbius 3-gonal (triangular) shape number with $A\left(2^k\right) = \frac{1}{2}\left(2^k - 1\right)$ by Theorem 3.*

Theorem 4. *Let n, m and m' be positive integers with greater than 1 and let $Ord_2(n) = m$ and $C(n) + 1 = m'$. Then, $A(n), B(n) \in \mathbb{Z}$ if and only if $n \equiv 1 \pmod 2$. Furthermore,*

$$A(n) = \sum_{k=1}^{m-1} U_k(n) + \frac{1}{2}(1+n) - m \tag{4}$$

and

$$B(n) = \sum_{k=1}^{m'-1} \varphi_k(n) + \frac{1}{2}(1+n) - m'. \tag{5}$$

Proof. First, we consider $A(n)$. We find the set $\{(0, U_0(n)), (1, U_1(n)), \ldots, (m, U_m(n))\}$. Thus, we have

$$A(n) = \frac{1}{2}\left(U_0(n) + U_1(n)\right) + \frac{1}{2}\left(U_1(n) + U_2(n)\right) + \ldots + \frac{1}{2}\left(U_{m-1}(n) + U_m(n)\right) - m$$

$$= U_1(n) + \ldots + U_{m-1}(n) + \frac{1}{2}(1+n) - m.$$

$$\equiv \frac{1}{2}(1+n) \pmod 1.$$

Similarly, we get (5). These complete the proof of Theorem 4. $\square$

3. Classification of the Absolute Möbius Divisor Function $U(n)$ with $Ord_2(n) = 2$

In this section, we study integers n when $Ord_2(n) = 2$. If $Ord_2(n) = 2$, then n has three cases which are 3-gonal (triangular) shape number, 4-gonal convex shape number, and 4-gonal non-convex shape numbers in Figure 5.

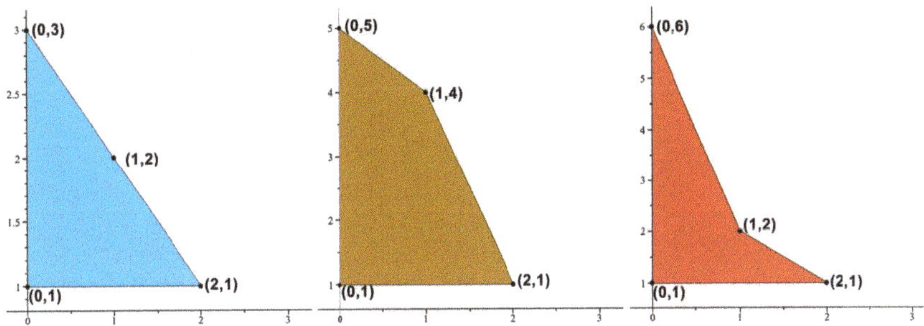

Figure 5. 3-gonal (triangular), 4-gonal convex, 4-gonal non-convex shapes.

Theorem 5. *Let $p_1, \ldots, p_r$ be Fermat primes and $e_1, \ldots, e_r$ be positive integers. If $n = 2^k p_1^{e_1} p_2^{e_2} \ldots p_r^{e_r}$, then $Ord_2(n) = 2$.*

Proof. Let

$$p_i = 2^{2^{m_i}} + 1 \quad (1 \leq i \leq r) \tag{6}$$

be Fermat primes. By Corollary 1 and Lemma 2 we have

$$
\begin{aligned}
U(n) &= U\left(2^k p_1^{e_1} p_2^{e_2} ... p_r^{e_r}\right) \\
&= U\left(2^k\right) U\left(p_1^{e_1}\right) U\left(p_2^{e_2}\right) ... U\left(p_r^{e_r}\right) \\
&= (p_1 - 1)(p_2 - 1) ... (p_r - 1) \\
&= 2^{2^{m_1}} 2^{2^{m_2}} ... 2^{2^{m_r}} \\
&= 2^t.
\end{aligned}
$$

Thus, we can see that $U(n) = U_1(n) = 2^t$ and $U_2(n) = U(U_1(n)) = U(2^t) = 1$. Therefore, we get Theorem 5. $\square$

The First 32 values of $U(n)$ and $\varphi(n)$ for $n = 2^k p_1^{e_1} p_2^{e_2} ... p_r^{e_r}$ are given by Table A2 (see Appendix B).

Remark 7. *Iterations of the odd divisor function $S(n)$, the absolute Möbius divisor function $U(n)$, and Euler totient function $\varphi(n)$ have small different properties. Table 6. gives an example of differences of $\varphi_k(n)$, $U_k(n)$, and $S_k(n)$ with $k = 1, 2$.*

Table 6. $\varphi_k(n)$, $U_k(n)$, and $S_k(n)$ with $k = 1, 2$.

Function f	$U(n)$	$\varphi(n)$	$S(n)$
$f_1(n) = 1$	$n = 2^k$	$n = 2^k$	$n = 2^k$
	$(k \geq 0)$	$(k = 0, 1)$	$(k \geq 0)$
$f_2(n) = 1$	$n = 2^k p_1^{e_1} ... p_r^{e_r}$	$n = 2^{k_1} 3^{k_2}$	$n = 2^k q_1 ... q_s$
	$(k \geq 0)$	$(k_1 = 0, 1)$	$(k \geq 0)$
	$p_i :$ Fermat primes	$(k_2 = 0, 1)$	$q_i :$ Mersenne primes
	(Theorem 5)	([2], p. 21)	([3], p. 3)

Lemma 3. *Let $n = p_i$ be Fermat primes. Then 3 is an absolute Möbius 3-gonal (triangular) shape number and $p_i (\neq 3)$ are absolute Möbius 4-gonal convex numbers.*

Proof. The set $\{(0, 3), (1, 2), (2, 1), (1, 0)\}$ makes a triangle. Let $p_i = 2^{2^{m_i}} + 1$ be a Fermat primes except 3. We get $U(p_i) = 2^{2^{m_i}}$. So, we get

$$
\mathbf{A} = \left\{ \left(0, 2^{2^{m_i}} + 1\right), \left(1, 2^{2^{m_i}}\right), (2, 1), (0, 1) \right\}.
$$

Because of $\left(2^{2^{m_i}} + 1 - 2^{2^{m_i}}\right) < \left(2^{2^{m_i}} - 1\right)$, the set $\mathbf{A}$ gives a convex shape. This completes the proof Lemma 3. $\square$

Lemma 4. *Let p_i be Fermat primes. Then $2^{m_1} p_i$ and $p_i^{m_2}$ are absolute Möbius 4-gonal non-convex shape numbers with $m_1, m_2 (\geq 2)$ positive integers.*

Proof. Let $p_i = 2^{2^{m_i}} + 1$ be a Fermat primes. Consider

$$
2^{m_1} p_i - (p_i - 1) = 2^{m_1} \cdot 2^{2^{m_i}} - 2^{2^{m_i}} \quad \text{and} \quad (p_i - 1) - 1 = 2^{2^{m_i}} - 1.
$$

So, $2^{m_1} p_i - (p_i - 1) > (p_i - 1) - 1$. Thus, $2^{m_1} p_i$ are absolute Möbius 4-gonal non-convex shape numbers. Similarly, we get $p_i^{m_1} - (p_i - 1) > (p_i - 1) - 1$.

Thus, these complete the proof Lemma 4. $\square$

Lemma 5. *Let $p_1, \ldots, p_r$ be Fermat primes. Then $2p_1 \ldots p_r$ are absolute Möbius 4-gonal non-convex shape numbers.*

Furthermore, if $m, e_1, \ldots e_r$ are positive integers then $2^m p_1^{e_1} \ldots p_r^{e_r}$ are absolute Möbius 4-gonal non-convex shape numbers.

Proof. The proof is similar to Lemma 4. $\square$

Lemma 6. *Let r be a positive integer. Then*

$$\prod_{i=0}^{r} \left(2^{2^i} + 1 \right) - 2 \prod_{i=0}^{r} 2^{2^i} + 1 = 0.$$

Proof. We note that

$$\prod_{i=0}^{r} \left(x^{2^i} + 1 \right) = \frac{x^{2^{r+1}} - 1}{x - 1} \text{ and } \prod_{i=0}^{r} x^{2^i} = x^{2^{r+1}-1}.$$

Let $f(x) := \prod_{i=0}^{r} \left(2^{2^i} + 1 \right) - 2 \prod_{i=0}^{r} 2^{2^i} + 1$. Thus $f(2) = 0$. This is completed the proof of Lemma 6. $\square$

Corollary 4. *Let $f_i \in F_1$. Then f_i is an absolute Möbius 3-gonal (triangular) shape number.*

Proof. It is trivial by Lemma 6. $\square$

Remark 8. *Fermat first conjectured that all the numbers in the form of $f_n = 2^{2^n} + 1$ are primes [16]. Up-to-date there are only five known Fermat primes. That is, $f_0 = 3$, $f_1 = 5$, $f_2 = 17$, $f_3 = 257$, and $f_4 = 65537$.*

Though we find a new Fermat prime p_6, 6th Fermat primes, we cannot find a new absolute Möbius 3-gonal (triangular) number by

$$\prod_{i=0}^{4} \left(2^{2^i} + 1 \right) \times \left(2^{2^{r'}} + 1 \right) - 2 \left(\prod_{i=0}^{4} 2^{2^i} \right) 2^{2^{r'}} + 1 > 0. \tag{7}$$

Lemma 7. *Let $p_1, p_2, \ldots, p_r, p_t$ be Fermat primes with $p_1 < p_2 < \ldots < p_r < p_t$ and $t > 5$. If $n = \prod_{i=1}^{r} p_i \in F_1$ then $n \times p_t$ are absolute Möbius 4-gonal convex shape numbers.*

Proof. Let $p_t = 2^{2^k} + 1$ be a Fermat prime, where k is a positive integer. We note that $r \leq 5$ and $p_t = 2^{2^k} + 1 > 2^{2^6} + 1$. In a similar way in (7), we obtain

$$p_1 \ldots p_r p_t - 2 \left(p_1 - 1 \right) \ldots \left(p_r - 1 \right) \left(p_t - 1 \right) + 1 =$$
$$\left\{ \prod_{i=0}^{r-1} \left(2^{2^i} + 1 \right) \right\} \left(2^{2^k} - 1 \right) - 2^{1+2^0+2^1+\ldots+2^{r-1}+2^k} + 1 > 0. \tag{8}$$

By Theorem 5, $Ord_2(n \times p_t) = 2$. By (8), $n \times p_t$ is an absolute Möbius 4-gonal convex shape number. This completes the proof of Lemma 7. $\square$

Lemma 8. *Let $p_1, p_2, \ldots, p_r, p_t$ be Fermat primes with $p_1 < p_2 < \ldots < p_r < p_t$.*

Then $m = p_1^{f_1} \cdots p_u^{f_u}$ are absolute Möbius 4-gonal non-convex shape numbers except $m \in F_0 \cup F_1 \cup F_2$.

Proof. Similar to Lemmas 5 and 7. $\square$

Proof of Theorem 1 (Main Theorem). It is completed by Remark 6, Theorem 5, Lemmas 3 and 4, Corollary 4, Remark 8, Lemmas 7 and 8. $\square$

Remark 9. *If n are absolute Möbius 3-gonal (triangular) or 4-gonal convex shape numbers then n is the regular n-gon by Gauss Theorem.*

Example 2. *The set V_3 is $\{(0,3),(1,2),(2,1),(0,1)\}$. Thus, a positive integer 3 is an absolute Möbius 3-gonal convex shape number.*

Similarly, 15, 255, 65535, 4294967295 are absolute Möbius 3-gonal convex numbers derived from

$$V_{15} = \{(0,15),(1,8),(2,1),(0,1)\},$$
$$V_{255} = \{(0,255),(1,128),(2,1),(0,1)\},$$
$$V_{65535} = \{(0,65535),(1,32768),(2,1),(0,1)\},$$
$$V_{4294967295} = \{(0,4294967295),(1,2147483648),(2,1),(0,1)\}.$$

Remark 10. *Let $Min(m)$ denote the minimal number of m-gonal number. By using Maple 13 Program, Table 7 shows us minimal numbers $Min(m)$ about from 3-gonal (triangular) to 14-gonal shape number.*

Table 7. Values of Min(m).

m	Min(m)	Prime or Not	m	Min(m)	Prime or Not
3	2	prime	9	719	prime
4	5	prime	10	1439	prime
5	7	prime	11	2879	prime
6	23	prime	12	34,549	prime
7	47	prime	13	138,197	prime
8	283	prime	14	1,266,767	prime

Conjecture 1. *For any positive integer $m \, (\geq 3)$, $Min(m)$ is a prime integer.*

Author Contributions: The definitions, lemmas, theorems and remarks within the paper are contributed by D.K., U.S. and S.I. Also, the introduction, body and conclusion sections are written by D.K., U.S. and S.I. The authors read and approved the final manuscript.

Funding: This work was funded by "Research Base Construction Fund Support Program" Jeonbuk National University in 2019. Supported by Balikesir University Research, G. No: 2017/20.

Conflicts of Interest: The authors declare that there is no conflict of interest.

Appendix A. Values of $U(n)$

Table A1. Values of $U(n)$ $(1 \leq n \leq 100)$.

n	1	2	3	4	5	6	7	8	9	10	11	12	13	14	15	16	17	18	19	20
$U(n)$	1	1	2	1	4	2	6	1	2	4	10	2	12	6	8	1	16	2	18	4
n	21	22	23	24	25	26	27	28	29	30	31	32	33	34	35	36	37	38	39	40
$U(n)$	12	10	22	2	4	12	2	6	28	8	30	1	20	16	24	2	36	18	24	4
n	41	42	43	44	45	46	47	48	49	50	51	52	53	54	55	56	57	58	59	60
$U(n)$	40	12	42	10	8	22	46	2	6	4	32	12	52	2	40	6	36	28	58	8
n	61	62	63	64	65	66	67	68	69	70	71	72	73	74	75	76	77	78	79	80
$U(n)$	60	30	12	1	48	20	66	16	44	24	70	2	72	36	8	18	60	24	78	4
n	81	82	83	84	85	86	87	88	89	90	91	92	93	94	95	96	97	98	99	100
$U(n)$	2	40	82	12	64	42	56	10	88	8	72	22	60	46	72	2	96	6	20	4

Appendix B. Values of $n = 2^k p_1 p_2 ... p_i$, $U(n)$, $\varphi(n)$

Table A2. Values of $n = 2^k p_1 p_2 ... p_i$, $U(n)$, $\varphi(n)$ with $Ord_2(n) = 2$.

n	$U(n)$	$\varphi(n)$	n	$U(n)$	$\varphi(n)$
3	2	2	$40 = 2^3 \times 5$	$4 = 2^2$	$16 = 2^4$
5	$4 = 2^2$	$4 = 2^2$	$45 = 3^2 \times 5$	$8 = 2^3$	$24 = 2^3 \times 3$
$6 = 2 \times 3$	2	2	$48 = 2^4 \times 3$	2	$16 = 2^4$
$9 = 3^2$	2	$6 = 2 \times 3$	$50 = 2 \times 5^2$	$4 = 2^2$	$20 = 2^4 \times 5$
$10 = 2 \times 5$	$4 = 2^2$	$4 = 2^2$	$51 = 3 \times 17$	$32 = 2^5$	$32 = 2^5$
$12 = 2^2 \times 3$	2	$4 = 2^2$	$54 = 2 \times 3^3$	2	$18 = 2 \times 3^2$
$15 = 3 \times 5$	$8 = 2^3$	$8 = 2^3$	$60 = 2^2 \times 3 \times 5$	$8 = 2^3$	$16 = 2^4$
17	$16 = 2^4$	$16 = 2^4$	$68 = 2^2 \times 17$	$16 = 2^4$	$32 = 2^5$
$18 = 2 \times 3^2$	2	$6 = 2 \times 3$	$72 = 2^3 \times 3^2$	2	$24 = 2^3 \times 3$
$20 = 2^2 \times 5$	$4 = 2^2$	$8 = 2^3$	$75 = 3 \times 5^2$	$8 = 2^3$	$40 = 2^3 \times 5$
$24 = 2^3 \times 3$	2	$8 = 2^3$	$80 = 2^4 \times 5$	$4 = 2^2$	$32 = 2^5$
$25 = 5^2$	$4 = 2^2$	$20 = 2^2 \times 5$	$81 = 3^4$	2	$54 = 2 \times 3^3$
$27 = 3^3$	2	$18 = 2 \times 3^2$	$85 = 5 \times 17$	$64 = 2^6$	$64 = 2^6$
$30 = 2 \times 3 \times 5$	$8 = 2^3$	$8 = 2^3$	$90 = 2 \times 3^2 \times 5$	$8 = 2^3$	$24 = 2^3 \times 3$
$34 = 2 \times 17$	$16 = 2^4$	$16 = 2^4$	$96 = 2^5 \times 3$	2	$32 = 2^5$
$36 = 2^2 \times 3^2$	2	$12 = 2^2 \times 3$	$100 = 2^2 \times 5^2$	$4 = 2^2$	$40 = 2^3 \times 5$

References

1. Rose, H.E. *A Course in Number Theory*; Clarendon Press: London, UK, 1994.
2. Shapiro, H. An analytic function arising from the φ function. *Am. Math. Mon.* **1943**, *50*, 18–30.
3. Kim, D.; Bayad, A. Polygon Numbers Associated with the Sum of Odd Divisors Function. *Exp. Math.* **2017**, *26*, 287–297. [CrossRef]
4. Dickson, L.E. *History of the Theory of Numbers. Vol. I: Divisibility and Primality*; Chelsea Pub. Co.: New York, NY, USA, 1966.
5. Erdös, P. Some Remarks on Euler's ϕ Function and Some Related Problems. *Bull. Am. Math. Soc.* **1945**, *51*, 540–544. [CrossRef]

6. Guy, R.K. *Unsolved Problems in Number Theory*; Springer: New York, NY, USA, 2004.

7. Ireland, K.; Rosen, M. *A Classical Introduction to Modern Number Theory*; GTM 84; Springer-Verlag: New York, NY, USA, 1990.

8. Moser, L. Some equations involving Eulers totient function. *Am. Math. Mon.* **1949**, *56*, 22–23. [CrossRef]

9. Sierpiński, W. *Elementary Theory of Numbers*; Polska Akademia Nauk: Warsaw, Poland, 1964.

10. Kaczorowski, J. On a generalization of the Euler totient function. *Monatshefte für Mathematik* **2012**, *170*, 27–48. [CrossRef]

11. Rassias, M.T. From a cotangent sum to a generalized totient function. *Appl. Anal. Discret. Math.* **2017**, *11*, 369–385. [CrossRef]

12. Shanks, D. *Solved and Unsolved Problems in Number Theory*; AMS: Providence, RI, USA, 2001.

13. Iwaniec, H.; Kowalski, E. *Analytic Number Theory*; American Mathematical Society: Providence, RI, USA, 2004; Volume 53.

14. Koblitz, N. *Introduction to Elliptic Curves and Modular Forms*; Springer-Verlag: New York, NY, USA, 1984.

15. Erdös, P.; Granville, A.; Pomerance, C.; Spiro, C. On the normal behavior of the iterates of some arithmetic functions. In *Analytic Number Theory*; Birkhäuser: Boston, MA, USA, 1990.

16. Tsang, C. Fermat Numbers. University of Washington. Available online: http://wstein.org/edu/2010/414/projects/tsang.pdf (accessed on 5 August 2019).

Article

Differential Equations Arising from the Generating Function of the (r, β)-Bell Polynomials and Distribution of Zeros of Equations

Kyung-Won Hwang [1], Cheon Seoung Ryoo [2,*] and Nam Soon Jung [3]

[1] Department of Mathematics, Dong-A University, Busan 604-714, Korea
[2] Department of Mathematics, Hannam University, Daejeon 34430, Korea
[3] College of Talmage Liberal Arts, Hannam University, Daejeon 34430, Korea
* Correspondence: ryoocs@hnu.kr

Received: 2 July 2019; Accepted: 9 August 2019; Published: 12 August 2019

Abstract: In this paper, we study differential equations arising from the generating function of the (r, β)-Bell polynomials. We give explicit identities for the (r, β)-Bell polynomials. Finally, we find the zeros of the (r, β)-Bell equations with numerical experiments.

Keywords: differential equations; Bell polynomials; r-Bell polynomials; (r, β)-Bell polynomials; zeros

MSC: 05A19; 11B83; 34A30; 65L99

---

1. Introduction

The moments of the Poisson distribution are a well-known connecting tool between Bell numbers and Stirling numbers. As we know, the Bell numbers B_n are those using generating function

$$e^{(e^t-1)} = \sum_{n=0}^{\infty} B_n \frac{t^n}{n!}.$$

The Bell polynomials $B_n(\lambda)$ are this formula using the generating function

$$e^{\lambda(e^t-1)} = \sum_{n=0}^{\infty} B_n(\lambda) \frac{t^n}{n!}, \tag{1}$$

(see [1,2]).

Observe that

$$B_n(\lambda) = \sum_{i=0}^{n} \lambda^i S_2(n, i),$$

where $S_2(n, i) = \frac{1}{i!} \sum_{l=0}^{i} (-1)^{i-l} \binom{i}{l} l^n$ denotes the second kind Stirling number.

The generalized Bell polynomials $B_n(x, \lambda)$ are these formula using the generating function:

$$\sum_{n=0}^{\infty} B_n(x, \lambda) \frac{t^n}{n!} = e^{xt - \lambda(e^t - t - 1)}, \text{ (see [2])}.$$

In particular, the generalized Bell polynomials $B_n(x, -\lambda) = E_\lambda[(Z + x - \lambda)^n], \lambda, x \in \mathbb{R}, n \in \mathbb{N}$, where Z is a Poission random variable with parameter $\lambda > 0$ (see [1–3]). The (r, β)-Bell polynomials $G_n(x, r, \beta)$ are this formula using the generating function:

$$F(t, x, r, \beta) = \sum_{n=0}^{\infty} G_n(x, r, \beta) \frac{t^n}{n!} = e^{rt + (e^{\beta t} - 1)\frac{x}{\beta}}, \tag{2}$$

(see [3]), where, β and r are real or complex numbers and $(r, \beta) \neq (0,0)$. Note that $B_n(x + r, -x) = G_n(x, r, 1)$ and $B_n(x) = G_n(x, 0, 1)$. The first few examples of (r, β)-Bell polynomials $G_n(x, r, \beta)$ are

$$G_0(x, r, \beta) = 1,$$
$$G_1(x, r, \beta) = r + x,$$
$$G_2(x, r, \beta) = r^2 + \beta x + 2rx + x^2,$$
$$G_3(x, r, \beta) = r^3 + \beta^2 x + 3\beta rx + 3r^2 x + 3\beta x^2 + 3rx^2 + x^3,$$
$$G_4(x, r, \beta) = r^4 + \beta^3 x + 4\beta^2 rx + 6\beta r^2 x + 4r^3 x + 7\beta^2 x^2 + 12\beta rx^2$$
$$\qquad + 6r^2 x^2 + 6\beta x^3 + 4rx^3 + x^4,$$
$$G_5(x, r, \beta) = r^5 + \beta^4 x + 5\beta^3 rx + 10\beta^2 r^2 x + 10\beta r^3 x + 5r^4 x + 15\beta^3 x^2 + 35\beta^2 rx^2$$
$$\qquad + 30\beta r^2 x^2 + 10r^3 x^2 + 25\beta^2 x^3 + 30\beta rx^3 + 10r^2 x^3 + 10\beta x^4 + 5rx^4 + x^5.$$

From (1) and (2), we see that

$$\sum_{n=0}^{\infty} G_n(x, r, \beta) \frac{t^n}{n!} = e^{(e^{\beta t} - 1)\frac{x}{\beta}} e^{rt}$$

$$= \left(\sum_{k=0}^{\infty} B_k(x/\beta)\beta^k \frac{t^k}{k!} \right) \left(\sum_{m=0}^{\infty} r^m \frac{t^m}{m!} \right) \tag{3}$$

$$= \sum_{n=0}^{\infty} \left(\sum_{k=0}^{n} \binom{n}{k} B_k(x/\beta)\beta^k r^{n-k} \right) \frac{t^n}{n!}.$$

Compare the coefficients in Formula (3). We can get

$$G_n(x, r, \beta) = \sum_{k=0}^{n} \binom{n}{k} \beta^k B_k(x/\beta) r^{n-k}, \quad (n \geq 0).$$

Similarly we also have

$$G_n(x + y, r, \beta) = \sum_{k=0}^{n} \binom{n}{k} G_k(x, r, \beta) B_{n-k}(y/\beta)\beta^{n-l}.$$

Recently, many mathematicians have studied the differential equations arising from the generating functions of special polynomials (see [4–8]). Inspired by their work, we give a differential equations by generation of (r, β)-Bell polynomials $G_n(x, r, \beta)$ as follows. Let D denote differentiation with respect to t, D^2 denote differentiation twice with respect to t, and so on; that is, for positive integer N,

$$D^N F = \left(\frac{\partial}{\partial t} \right)^N F(t, x, r, \beta).$$

We find differential equations with coefficients $a_i(N, x, r, \beta)$, which are satisfied by

$$\left(\frac{\partial}{\partial t} \right)^N F(t, x, r, \beta) - a_0(N, x, r, \beta) F(t, x, r, \beta) - \cdots - a_N(N, x, r, \beta) e^{\beta t N} F(t, x, r, \beta) = 0.$$

Using the coefficients of this differential equation, we give explicit identities for the (r, β)-Bell polynomials. In addition, we investigate the zeros of the (r, β)-Bell equations with numerical methods. Finally, we observe an interesting phenomena of 'scattering' of the zeros of (r, β)-Bell equations. Conjectures are also presented through numerical experiments.

2. Differential Equations Related to (R, β)-Bell Polynomials

Differential equations arising from the generating functions of special polynomials are studied by many authors to give explicit identities for special polynomials (see [4–8]). In this section, we study differential equations arising from the generating functions of (r, β)-Bell polynomials.

Let

$$F = F(t, x, r, \beta) = \sum_{n=0}^{\infty} G_n(x, r, \beta) \frac{t^n}{n!} = e^{rt + (e^{\beta t} - 1)\frac{x}{\beta}}, \quad x, r, \beta \in \mathbb{C}. \tag{4}$$

Then, by (4), we have

$$
\begin{aligned}
DF = \frac{\partial}{\partial t} F(t, x, r, \beta) &= \frac{\partial}{\partial t} \left(e^{rt + (e^{\beta t} - 1)\frac{x}{\beta}} \right) \\
&= e^{rt + (e^{\beta t} - 1)\frac{x}{\beta}} (r + xe^{\beta t}) \\
&= re^{rt + (e^{\beta t} - 1)\frac{x}{\beta}} + xe^{(r+\beta)t + (e^{\beta t} - 1)\frac{x}{\beta}} \\
&= rF(t, x, r, \beta) + xF(t, x, r + \beta, \beta),
\end{aligned}
\tag{5}
$$

$$
\begin{aligned}
D^2 F &= rDF(t, x, r, \beta) + xDF(t, x, r + \beta, \beta) \\
&= r^2 F(t, x, r, \beta) + x(2r + \beta)F(t, x, r + \beta, \beta) + x^2 F(t, x, r + 2\beta, \beta),
\end{aligned}
\tag{6}
$$

and

$$
\begin{aligned}
D^3 F &= r^2 DF(t, x, r, \beta) + x(2r + \beta)DF(t, x, r + \beta, \beta) + x^2 DF(t, x, r + 2\beta, \beta) \\
&= r^3 F(t, x, r, \beta) + x \left(r^2 + (2r + \beta)(r + \beta) \right) F(t, x, r + \beta, \beta) \\
&\quad + x^2 (3r + 3\beta)F(t, x, r + 2\beta, \beta) + x^3 F(t, x, r + 3\beta, \beta).
\end{aligned}
$$

We prove this process by induction. Suppose that

$$D^N F = \sum_{i=0}^{N} a_i(N, x, r, \beta)F(t, x, r + i\beta, \beta), (N = 0, 1, 2, \ldots). \tag{7}$$

is true for N. From (7), we get

$$
\begin{aligned}
D^{N+1} F &= \sum_{i=0}^{N} a_i(N, x, r, \beta)DF(t, x, r + i\beta, \beta) \\
&= \sum_{i=0}^{N} a_i(N, x, r, \beta) \left\{ (r + i\beta)F(t, x, r + i\beta, \beta) + xF(t, x, r + (i+1)\beta, \beta) \right\} \\
&= \sum_{i=0}^{N} a_i(N, x, r, \beta)(r + i\beta)F(t, x, r + i\beta, \beta) \\
&\quad + x \sum_{i=0}^{N} a_i(N, x, r, \beta)F(t, x, r + (i+1)\beta, \beta) \\
&= \sum_{i=0}^{N} (r + i\beta)a_i(N, x, r, \beta)F(t, x, r + i\beta, \beta) \\
&\quad + x \sum_{i=1}^{N+1} a_{i-1}(N, x, r, \beta)F(t, x, r + i\beta, \beta).
\end{aligned}
\tag{8}
$$

From (8), we get

$$D^{N+1} F = \sum_{i=0}^{N+1} a_i(N + 1, x, r, \beta)F(t, x, r + i\beta, \beta). \tag{9}$$

We prove that

$$D^{k+1}F = \sum_{i=0}^{k+1} a_i(k+1, x, r, \beta) F(t, x, r+i\beta, \beta).$$

If we compare the coefficients on both sides of (8) and (9), then we get

$$a_0(N+1, x, r, \beta) = ra_0(N, x, r, \beta), \quad a_{N+1}(N+1, x, r, \beta) = xa_N(N, x, r, \beta), \tag{10}$$

and

$$a_i(N+1, x, r, \beta) = (r+i\beta)a_{i-1}(N, x, r, \beta) + xa_{i-1}(N, x, r, \beta), (1 \le i \le N). \tag{11}$$

In addition, we get

$$F(t, x, r, \beta) = a_0(0, x, r, \beta)F(t, x, r, \beta). \tag{12}$$

Now, by (10), (11) and (12), we can obtain the coefficients $a_i(j, x, r, \beta)_{0 \le i, j \le N+1}$ as follows. By (12), we get

$$a_0(0, x, r, \beta) = 1. \tag{13}$$

It is not difficult to show that

$$
\begin{aligned}
rF(t, x, r, \beta) &+ xF(t, x, r+\beta, \beta) \\
&= DF(t, x, r, \beta) \\
&= \sum_{i=0}^{1} a_i(1, x, r, \beta)F(t, x, r+\beta, \beta) \\
&= a_0(1, x, r, \beta)F(t, x, r, \beta) + a_1(1, x, r, \beta)F(t, x, r+\beta, \beta).
\end{aligned}
\tag{14}
$$

Thus, by (14), we also get

$$a_0(1, x, r, \beta) = r, \quad a_1(1, x, r, \beta) = x. \tag{15}$$

From (10), we have that

$$a_0(N+1, x, r, \beta) = ra_0(N, x, r, \beta) = \cdots = r^N a_0(1, x, r, \beta) = r^{N+1}, \tag{16}$$

and

$$a_{N+1}(N+1, x, r, \beta) = xa_N(N, x, r, \beta) = \cdots = x^N a_1(1, x, r, \beta) = x^{N+1}. \tag{17}$$

For $i = 1, 2, 3$ in (11), we have

$$a_1(N+1, x, r, \beta) = x \sum_{k=0}^{N} (r+\beta)^k a_0(N-k, x, r, \beta), \tag{18}$$

$$a_2(N+1, x, r, \beta) = x \sum_{k=0}^{N-1} (r+2\beta)^k a_1(N-k, x, r, \beta), \tag{19}$$

and

$$a_3(N+1, x, r, \beta) = x \sum_{k=0}^{N-2} (r+3\beta)^k a_2(N-k, x, r, \beta). \tag{20}$$

By induction on i, we can easily prove that, for $1 \le i \le N$,

$$a_i(N+1, x, r, \beta) = x \sum_{k=0}^{N-i+1} (r+i\beta)^k a_{i-1}(N-k, x, r, \beta). \tag{21}$$

Here, we note that the matrix $a_i(j, x, r, \beta)_{0 \le i, j \le N+1}$ is given by

$$
\begin{pmatrix}
1 & r & r^2 & r^3 & \cdots & r^{N+1} \\
0 & x & x(2r + \beta) & x(3r^2 + 3r\beta + \beta^2) & \cdots & \cdot \\
0 & 0 & x^2 & x^2(3r + 3\beta) & \cdots & \cdot \\
0 & 0 & 0 & x^3 & \cdots & \cdot \\
\vdots & \vdots & \vdots & \vdots & \ddots & \vdots \\
0 & 0 & 0 & 0 & \cdots & x^{N+1}
\end{pmatrix}
$$

Now, we give explicit expressions for $a_i(N + 1, x, r, \beta)$. By (18), (19), and (20), we get

$$
a_1(N + 1, x, r, \beta) = x \sum_{k_1=0}^{N} (r + \beta)^{k_1} a_0(N - k_1, x, r, \beta)
$$

$$
= \sum_{k_1=0}^{N} (r + \beta)^{k_1} r^{N - k_1},
$$

$$
a_2(N + 1, x, r, \beta) = x \sum_{k_2=0}^{N-1} (r + 2\beta)^{k_2} a_1(N - k_2, x, r, \beta)
$$

$$
= x^2 \sum_{k_2=0}^{N-1} \sum_{k_1=0}^{N-1-k_2} (r + \beta)^{k_1} (r + 2\beta)^{k_2} r^{N - k_2 - k_1 - 1},
$$

and

$$
a_3(N + 1, x, r, \beta)
$$

$$
= x \sum_{k_3=0}^{N-2} (r + 3\beta)^{k_3} a_2(N - k_3, x, r, \beta)
$$

$$
= x^3 \sum_{k_3=0}^{N-2} \sum_{k_2=0}^{N-2-k_3} \sum_{k_1=0}^{N-2-k_3-k_2} (r + 3\beta)^{k_3} (r + 2\beta)^{k_2} (r + \beta)^{k_1} r^{N - k_3 - k_2 - k_1 - 2}.
$$

By induction on i, we have

$$
a_i(N + 1, x, r, \beta)
$$

$$
= x^i \sum_{k_i=0}^{N-i+1} \sum_{k_{i-1}=0}^{N-i+1-k_i} \cdots \sum_{k_1=0}^{N-i+1-k_i-\cdots-k_2} \left(\prod_{l=1}^{i} (r + l\beta)^{k_l} \right) r^{N-i+1-\sum_{l=1}^{i} k_l}. \tag{22}
$$

Finally, by (22), we can derive a differential equations with coefficients $a_i(N, x, r, \beta)$, which is satisfied by

$$
\left(\frac{\partial}{\partial t} \right)^N F(t, x, r, \beta) - a_0(N, x, r, \beta) F(t, x, r, \beta) - \cdots - a_N(N, x, r, \beta) e^{\beta t N} F(t, x, r, \beta) = 0.
$$

Theorem 1. *For same as below $N = 0, 1, 2, \ldots$, the differential equation*

$$
D^N F = \sum_{i=0}^{N} a_i(N, x, r, \beta) e^{i\beta t} F(t, x, r, \beta)
$$

has a solution

$$
F = F(t, x, r, \beta) = e^{rt + (e^{\beta t} - 1)\frac{x}{\beta}},
$$

where

$$a_0(N, x, r, \beta) = r^N,$$
$$a_N(N, x, r, \beta) = x^N,$$
$$a_i(N, x, r, \beta) = x^i \sum_{k_i=0}^{N-i} \sum_{k_{i-1}=0}^{N-i-k_i} \cdots \sum_{k_1=0}^{N-i-k_i-\cdots-k_2} \left(\prod_{l=1}^{i} (r + l\beta)^{k_l} \right) r^{N-i-\sum_{l=1}^{i} k_l},$$
$$(1 \le i \le N).$$

From (4), we have this

$$D^N F = \left(\frac{\partial}{\partial t} \right)^N F(t, x, r, \beta) = \sum_{k=0}^{\infty} G_{k+N}(x, r, \beta) \frac{t^k}{k!}. \tag{23}$$

By using Theorem 1 and (23), we can get this equation:

$$\begin{aligned}
\sum_{k=0}^{\infty} G_{k+N}(x, r, \beta) \frac{t^k}{k!} &= D^N F \\
&= \left(\sum_{i=0}^{N} a_i(N, x, r, \beta) e^{i\beta t} \right) F(t, x, r, \beta) \\
&= \sum_{i=0}^{N} a_i(N, x, r, \beta) \left(\sum_{l=0}^{\infty} (i\beta)^l \frac{t^l}{l!} \right) \left(\sum_{m=0}^{\infty} G_m(x, r, \beta) \frac{t^m}{m!} \right) \\
&= \sum_{i=0}^{N} a_i(N, x, r, \beta) \left(\sum_{k=0}^{\infty} \sum_{m=0}^{k} \binom{k}{m} (i\beta)^{k-m} G_m(x, r, \beta) \frac{t^k}{k!} \right) \\
&= \sum_{k=0}^{\infty} \left(\sum_{i=0}^{N} \sum_{m=0}^{k} \binom{k}{m} (i\beta)^{k-m} a_i(N, x, r, \beta) G_m(x, r, \beta) \right) \frac{t^k}{k!}.
\end{aligned} \tag{24}$$

Compare coefficients in (24). We get the below theorem.

Theorem 2. *For* $k, N = 0, 1, 2, \ldots,$ *we have*

$$G_{k+N}(x, r, \beta) = \sum_{i=0}^{N} \sum_{m=0}^{k} \binom{k}{m} i^{k-m} \beta^{k-m} a_i(N, x, r, \beta) G_m(x, r, \beta), \tag{25}$$

where

$$a_0(N, x, r, \beta) = r^N,$$
$$a_N(N, x, r, \beta) = x^N,$$
$$a_i(N, x, r, \beta) = x^i \sum_{k_i=0}^{N-i} \sum_{k_{i-1}=0}^{N-i-k_i} \cdots \sum_{k_1=0}^{N-i-k_i-\cdots-k_2} \left(\prod_{l=1}^{i} (r + l\beta)^{k_l} \right) r^{N-i-\sum_{l=1}^{i} k_l},$$
$$(1 \le i \le N).$$

By using the coefficients of this differential equation, we give explicit identities for the (r, β)-Bell polynomials. That is, in (25) if $k = 0$, we have corollary.

Corollary 1. *For* $N = 0, 1, 2, \ldots,$ *we have*

$$G_N(x, r, \beta) = \sum_{i=0}^{N} a_i(N, x, r, \beta).$$

For $N = 0, 1, 2, \ldots$, it follows that equation

$$D^N F - \sum_{i=0}^{N} a_i(N, x, r, \beta) e^{i\beta t} F(t, x, r, \beta) = 0$$

has a solution

$$F = F(t, x, r, \beta) = e^{rt + (e^{\beta t} - 1)\frac{x}{\beta}}.$$

In Figure 1, we have a sketch of the surface about the solution F of this differential equation. On the left of Figure 1, we give $-3 \leq x \leq 3, -1 \leq t \leq 1$, and $r = 2, \beta = 5$. On the right of Figure 1, we give $-3 \leq x \leq 3, -1 \leq t \leq 1$, and $r = -3, \beta = 2$.

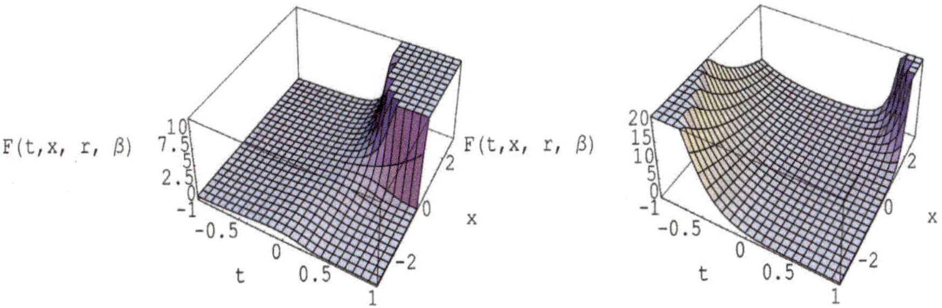

Figure 1. The surface for the solution $F(t, x, r, \beta)$.

Making N-times derivative for (4) with respect to t, we obtain

$$\left(\frac{\partial}{\partial t}\right)^N F(t, x, r, \beta) = \left(\frac{\partial}{\partial t}\right)^N e^{rt + (e^{\beta t} - 1)\frac{x}{\beta}} = \sum_{m=0}^{\infty} G_{m+N}(x, r, \beta) \frac{t^m}{m!}. \tag{26}$$

By multiplying the exponential series $e^{xt} = \sum_{m=0}^{\infty} x^m \dfrac{t^m}{m!}$ in both sides of (26) and Cauchy product, we derive

$$e^{-nt} \left(\frac{\partial}{\partial t}\right)^N F(t, x, r, \beta) = \left(\sum_{m=0}^{\infty} (-n)^m \frac{t^m}{m!}\right) \left(\sum_{m=0}^{\infty} G_{m+N}(x, r, \beta) \frac{t^m}{m!}\right)$$
$$= \sum_{m=0}^{\infty} \left(\sum_{k=0}^{m} \binom{m}{k} (-n)^{m-k} G_{N+k}(x, r, \beta)\right) \frac{t^m}{m!}. \tag{27}$$

By using the Leibniz rule and inverse relation, we obtain

$$e^{-nt} \left(\frac{\partial}{\partial t}\right)^N F(t, x, y) = \sum_{k=0}^{N} \binom{N}{k} n^{N-k} \left(\frac{\partial}{\partial t}\right)^k (e^{-nt} F(t, x, r, \beta))$$
$$= \sum_{m=0}^{\infty} \left(\sum_{k=0}^{N} \binom{N}{k} n^{N-k} G_{m+k}(x - n, r, \beta)\right) \frac{t^m}{m!}. \tag{28}$$

So using (27) and (28), and using the coefficients of $\dfrac{t^m}{m!}$ gives the below theorem.

Theorem 3. *Let m, n, N be nonnegative integers. Then*

$$\sum_{k=0}^{m} \binom{m}{k} (-n)^{m-k} G_{N+k}(x, r, \beta) = \sum_{k=0}^{N} \binom{N}{k} n^{N-k} G_{m+k}(x - n, r, \beta). \tag{29}$$

When we give $m = 0$ in (29), then we get corollary.

Corollary 2. *For $N = 0, 1, 2, \ldots$, we have*

$$G_N(x, r, \beta) = \sum_{k=0}^{N} \binom{N}{k} n^{N-k} G_k(x - n, r, \beta).$$

3. Distribution of Zeros of the (R, β)-Bell Equations

This section aims to demonstrate the benefit of using numerical investigation to support theoretical prediction and to discover new interesting patterns of the zeros of the (r, β)-Bell equations $G_n(x, r, \beta) = 0$. We investigate the zeros of the (r, β)-Bell equations $G_n(x, r, \beta) = 0$ with numerical experiments. We plot the zeros of the $B_n(x, \lambda) = 0$ for $n = 16, r = -5, -3, 3, 5, \beta = 2, 3$ and $x \in \mathbb{C}$ (Figure 2).

In top-left of Figure 2, we choose $n = 16$ and $r = -5, \beta = 2$. In top-right of Figure 2, we choose $n = 16$ and $r = -3, \beta = 3$. In bottom-left of Figure 2, we choose $n = 16$ and $r = 3, \beta = 2$. In bottom-right of Figure 2, we choose $n = 16$ and $r = 5, \beta = 3$.

Prove that $G_n(x, r, \beta), x \in \mathbb{C}$, has $Im(x) = 0$ reflection symmetry analytic complex functions (see Figure 3). Stacks of zeros of the (r, β)-Bell equations $G_n(x, r, \beta) = 0$ for $1 \leq n \leq 20$ from a 3-D structure are presented (Figure 3).

On the left of Figure 3, we choose $r = -5$ and $\beta = 2$. On the right of Figure 3, we choose $r = 5$ and $\beta = 2$. In Figure 3, the same color has the same degree n of (r, β)-Bell polynomials $G_n(x, r, \beta)$. For example, if $n = 20$, zeros of the (r, β)-Bell equations $G_n(x, r, \beta) = 0$ is red.

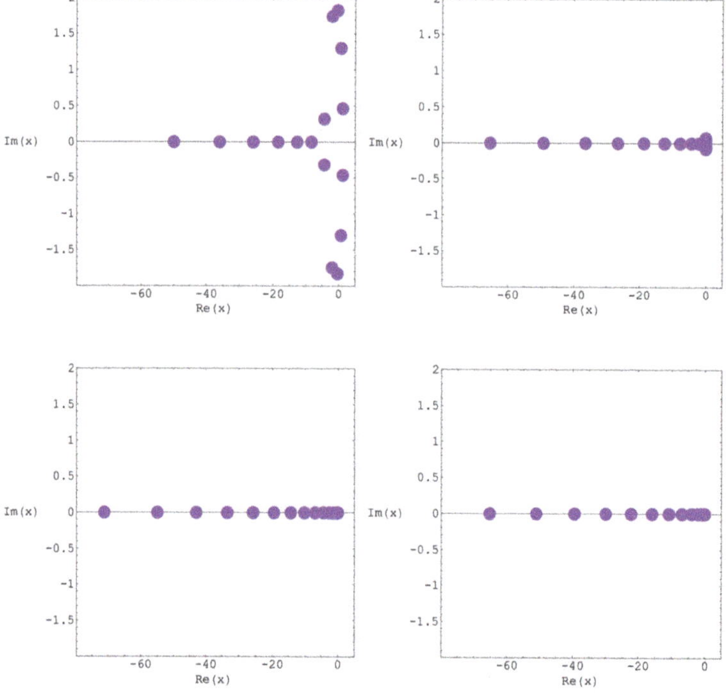

Figure 2. Zeros of $G_n(x, r, \beta) = 0$.

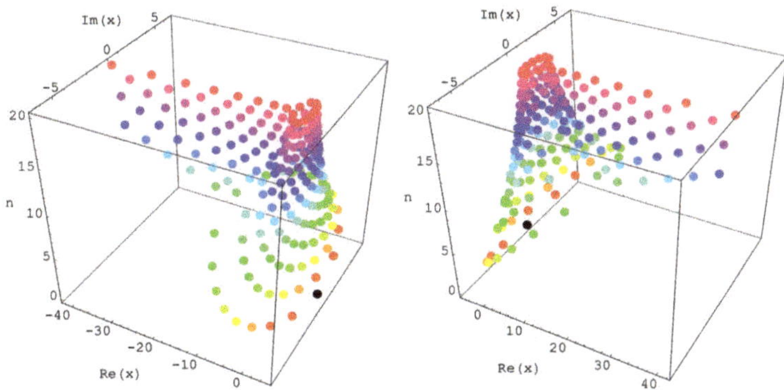

Figure 3. Stacks of zeros of $G_n(x, r, \beta) = 0, 1 \leq n \leq 20$.

Our numerical results for approximate solutions of real zeros of the (r, β)-Bell equations $G_n(x, r, \beta) = 0$ are displayed (Tables 1 and 2).

Table 1. Numbers of real and complex zeros of $G_n(x, r, \beta) = 0$

Degree n	$r = -5, \beta = 2$		$r = 5, \beta = 2$	
	Real Zeros	**Complex Zeros**	**Real Zeros**	**Xomplex Zeros**
1	1	0	1	0
2	0	2	2	0
3	1	2	3	0
4	0	4	4	0
5	1	4	5	0
6	0	6	6	0
7	1	6	7	0
8	0	8	8	0
9	1	8	9	0
10	2	8	10	0

Table 2. Approximate solutions of $G_n(x, r, \beta) = 0, x \in \mathbb{R}$.

Degree n	x
1	-5.000
2	$-9.317, \quad -2.683$
3	$-13.72, \quad -5.68, \quad -1.605$
4	$-18.21, \quad -9.01, \quad -3.77, \quad -1.010$
5	$-22.8, \quad -12.6, \quad -6.4, \quad -2.61, \quad -0.655$
6	$-27.4, \quad -16.3, \quad -9.3, \quad -4.7, \quad -1.85, \quad -0.434$
7	$-32.0, \quad -20.0, \quad -12.0, \quad -7.1, \quad -3.5, \quad -1.34, \quad -0.291$

Plot of real zeros of $G_n(x, r, \beta) = 0$ for $1 \leq n \leq 20$ structure are presented (Figure 4).

In Figure 4 (left), we choose $r = 5$ and $\beta = -2$. In Figure 4 (right), we choose $r = 5$ and $\beta = 2$. In Figure 4, the same color has the same degree n of (r, β)-Bell polynomials $G_n(x, r, \beta)$. For example, if $n = 20$, real zeros of the (r, β)-Bell equations $G_n(x, r, \beta) = 0$ is blue.

Next, we calculated an approximate solution satisfying $G_n(x, r, \beta) = 0, r = 5, \beta = 2, x \in \mathbb{R}$. The results are given in Table 2.

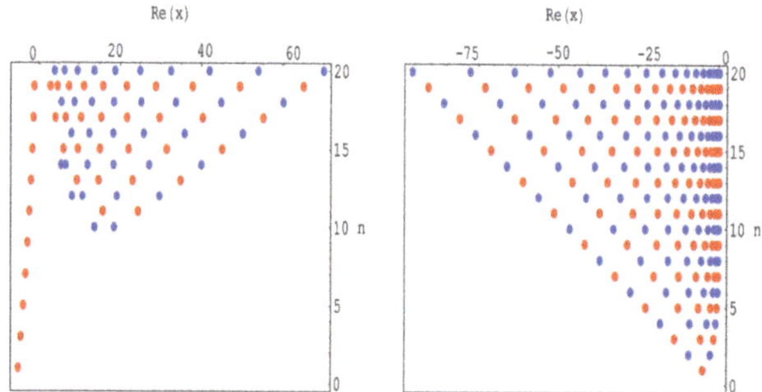

Figure 4. Stacks of zeros of $G_n(x, r, \beta) = 0, 1 \leq n \leq 20$.

4. Conclusions

We constructed differential equations arising from the generating function of the (r, β)-Bell polynomials. This study obtained the some explicit identities for (r, β)-Bell polynomials $G_n(x, r, \beta)$ using the coefficients of this differential equation. The distribution and symmetry of the roots of the (r, β)-Bell equations $G_n(x, r, \beta) = 0$ were investigated. We investigated the symmetry of the zeros of the (r, β)-Bell equations $G_n(x, r, \beta) = 0$ for various variables r and β, but, unfortunately, we could not find a regular pattern. We make the following series of conjectures with numerical experiments:

Let us use the following notations. $R_{G_n(x,r,\beta)}$ denotes the number of real zeros of $G_n(x, r, \beta) = 0$ lying on the real plane $Im(x) = 0$ and $C_{G_n(x,r,\beta)}$ denotes the number of complex zeros of $G_n(x, r, \beta) = 0$. Since n is the degree of the polynomial $G_n(x, r, \beta)$, we have $R_{G_n(x,r,\beta)} = n - C_{G_n(x,r,\beta)}$ (see Table 1).

We can see a good regular pattern of the complex roots of the (r, β)-Bell equations $G_n(x, r, \beta) = 0$ for $r > 0$ and $\beta > 0$. Therefore, the following conjecture is possible.

Conjecture 1. *For $r > 0$ and $\beta > 0$, prove or disprove that*

$$C_{H_n(x,y)} = 0.$$

As a result of investigating more $r > 0$ and $\beta > 0$ variables, it is still unknown whether the conjecture 1 is true or false for all variables $r > 0$ and $\beta > 0$ (see Figure 1 and Table 1).

We observe that solutions of (r, β)-Bell equations $G_n(x, r, \beta) = 0$ has $Im(x) = 0$, reflecting symmetry analytic complex functions. It is expected that solutions of (r, β)-Bell equations $G_n(x, r, \beta) = 0$, has not $Re(x) = a$ reflection symmetry for $a \in \mathbb{R}$ (see Figures 2–4).

Conjecture 2. *Prove or disprove that solutions of (r, β)-Bell equations $G_n(x, r, \beta) = 0$, has not $Re(x) = a$ reflection symmetry for $a \in \mathbb{R}$.*

Finally, how many zeros do $G_n(x, r, \beta) = 0$ have? We are not able to decide if $G_n(x, r, \beta) = 0$ has n distinct solutions (see Tables 1 and 2). We would like to know the number of complex zeros $C_{G_n(x,r,\beta)}$ of $G_n(x, r, \beta) = 0, Im(x) \neq 0$.

Conjecture 3. *Prove or disprove that $G_n(x, r, \beta) = 0$ has n distinct solutions.*

As a result of investigating more n variables, it is still unknown whether the conjecture is true or false for all variables n (see Tables 1 and 2). We expect that research in these directions will make a new approach using the numerical method related to the research of the (r, β)-Bell numbers and polynomials which appear in mathematics, applied mathematics, statistics, and mathematical physics. The reader may refer to [5–10] for the details.

Author Contributions: These authors contributed equally to this work.

Funding: This work was supported by the Dong-A university research fund.

Acknowledgments: The authors would like to thank the referees for their valuable comments, which improved the original manuscript in its present form.

Conflicts of Interest: The authors declare no conflicts of interest.

References

1. Mező, I. The r-Bell Numbers. *J. Integer. Seq.* **2010**, *13*. Available online: https://www.google.com.hk/url?sa=t&rct=j&q=&esrc=s&source=web&cd=1&ved=2ahUKEwiq6diglPfjAhW9yosBHX4lAo8QFjAAegQIBBAC&url=https%3A%2F%2Fcs.uwaterloo.ca%2Fjournals%2FJIS%2FVOL13%2FMezo%2Fmezo8.pdf&usg=AOvVaw0N25qEl3ROosJgHzsnxrlv (accessed on 10 July 2019).
2. Privault, N. Genrealized Bell polynomials and the combinatorics of Poisson central moments. *Electr. J. Comb.* **2011**, *18*, 54.
3. Corcino, R.B.; Corcino, C.B. On generalized Bell polynomials. *Discret. Dyn. Nat. Soc.* **2011**, *2011*. [CrossRef]
4. Kim, T.; Kim, D.S. Identities involving degenerate Euler numbers and polynomials arising from non-linear differential equations. *J. Nonlinear Sci. Appl.* **2016**, *9*, 2086–2098. [CrossRef]
5. Kim, T.; Kim, D.S.; Ryoo, C.S.; Kwon, H.I. Differential equations associated with Mahler and Sheffer-Mahler polynomials. *Nonlinear Funct. Anal. Appl.* **2019**, *24*, 453–462.
6. Ryoo, C.S.; Agarwal, R.P. ; Kang, J.Y. Differential equations arising from Bell-Carlitz polynomials and computation of their zeros. *Neural Parallel Sci. Comput.* **2016**, *24*, 93–107.
7. Ryoo, C.S. Differential equations associated with tangent numbers. *J. Appl. Math. Inf.* **2016**, *34*, 487–494. [CrossRef]
8. Ryoo, C.S. Differential equations associated with generalized Bell polynomials and their zeros. *Open Math.* **2016**, *14*, 807–815. [CrossRef]
9. Ryoo, C.S. A numerical investigation on the structure of the zeros of the degenerate Euler-tangent mixed-type polynomials. *J. Nonlinear Sci. Appl.* **2017**, *10*, 4474–4484. [CrossRef]
10. Ryoo, C.S.; Hwang, K.W.; Kim, D.J.; Jung, N.S. Dynamics of the zeros of analytic continued polynomials and differential equations associated with q-tangent polynomials. *J. Nonlinear Sci. Appl.* **2018**, *11*, 785–797. [CrossRef]

 *mathematics*

Article

Truncated Fubini Polynomials

Ugur Duran [1,*] **and Mehmet Acikgoz** [2]

[1] Department of the Basic Concepts of Engineering, Faculty of Engineering and Natural Sciences, Iskenderun Technical University, TR-31200 Hatay, Turkey
[2] Department of Mathematics, Faculty of Science and Arts, University of Gaziantep, TR-27310 Gaziantep, Turkey; acikgoz@gantep.edu.tr
* Correspondence: mtdrnugur@gmail.com

Received: 22 April 2019; Accepted: 9 May 2019; Published: 15 May 2019

Abstract: In this paper, we introduce the two-variable truncated Fubini polynomials and numbers and then investigate many relations and formulas for these polynomials and numbers, including summation formulas, recurrence relations, and the derivative property. We also give some formulas related to the truncated Stirling numbers of the second kind and Apostol-type Stirling numbers of the second kind. Moreover, we derive multifarious correlations associated with the truncated Euler polynomials and truncated Bernoulli polynomials.

Keywords: Fubini polynomials; Euler polynomials; Bernoulli polynomials; truncated exponential polynomials; Stirling numbers of the second kind

MSC: Primary 11B68; Secondary 11B83, 11B37, 05A19

---

1. Introduction

The classical Bernoulli and Euler polynomials are defined by means of the following generating functions:

$$\sum_{n=0}^{\infty} B_n\left(x\right) \frac{t^n}{n!} = \frac{t}{e^t - 1} e^{xt} \quad \left(|t| < 2\pi\right) \tag{1}$$

and:

$$\sum_{n=0}^{\infty} E_n\left(x\right) \frac{t^n}{n!} = \frac{2}{e^t + 1} e^{xt} \quad \left(|t| < \pi\right), \tag{2}$$

see [1–10] for details about the aforesaid polynomials. The Bernoulli numbers B_n and Euler numbers E_n are obtained by the special cases of the corresponding polynomials at $x = 0$, namely:

$$B_n\left(0\right) := B_n \text{ and } E_n\left(0\right) := E_n. \tag{3}$$

The truncated exponential polynomials have played a role of crucial importance to evaluate integrals including products of special functions; cf. [11], and also see the references cited therein. Recently, several mathematicians have studied truncated-type special polynomials such as truncated Bernoulli polynomials and truncated Euler polynomials; cf. [1,4,7,9,11,12].

Mathematics **2019**, *7*, 431; doi:10.3390/math7050431   www.mdpi.com/journal/mathematics

For non-negative integer m, the truncated Bernoulli and truncated Euler polynomials are introduced as follows:

$$\sum_{n=0}^{\infty} B_{m,n}(x) \frac{t^n}{n!} = \frac{\frac{t^m}{m!}}{e^t - \sum_{j=0}^{m-1} \frac{t^j}{j!}} e^{xt} \quad (\text{cf. } [1]) \tag{4}$$

and:

$$\sum_{n=0}^{\infty} E_{m,n}(x) \frac{t^n}{n!} = \frac{2\frac{t^m}{m!}}{e^t + 1 - \sum_{j=0}^{m-1} \frac{t^j}{j!}} e^{xt} \quad (\text{cf. } [7]). \tag{5}$$

Upon setting $x = 0$ in (4) and (5), the mentioned polynomials ($B_{m,n}(x)$ and $E_{m,n}(x)$), reduce to the corresponding numbers:

$$B_{m,n}(0) := B_{m,n} \text{ and } E_{m,n}(0) := E_{m,n} \tag{6}$$

termed as the truncated Bernoulli numbers and truncated Euler numbers, respectively.

Remark 1. *Setting $m = 0$ in (4) and $m = 1$ (5), then the truncated Bernoulli and truncated Euler polynomials reduce to the classical Bernoulli and Euler polynomials in (1) and (2).*

The Stirling numbers of the second kind are given by the following exponential generating function:

$$\sum_{n=0}^{\infty} S_2(n,k) \frac{t^n}{n!} = \frac{(e^t - 1)^k}{k!} \quad (\text{cf. } [2\text{--}5,7,8,10,13]) \tag{7}$$

or by the recurrence relation for a fixed non-negative integer ζ,

$$x^\zeta = \sum_{\mu=0}^{\zeta} S_2(\zeta, \mu)(x)_\mu, \tag{8}$$

where the notation $(x)_\mu$ called the falling factorial equals $x(x-1)\cdots(x-\mu+1)$; cf. [2–5,7–10,13], and see also the references cited therein.

The Apostol-type Stirling numbers of the second kind is defined by (cf. [8]):

$$\sum_{n=0}^{\infty} S_2(n,k:\lambda) \frac{t^n}{n!} = \frac{(\lambda e^t - 1)^k}{k!} \quad (\lambda \in \mathbb{C}/\{1\}). \tag{9}$$

The following sections are planned as follows: the second section includes the definition of the two-variable truncated Fubini polynomials and provides several formulas and relations including Stirling numbers of the second kind with several extensions. The third part covers the correlations for the two-variable truncated Fubini polynomials associated with the truncated Euler polynomials and the truncated Bernoulli polynomials. The last part of this paper analyzes the results acquired in this paper.

2. Two-Variable Truncated Fubini Polynomials

In this part, we define the two-variable truncated Fubini polynomials and numbers. We investigate several relations and identities for these polynomials and numbers.

We firstly remember the classical two-variable Fubini polynomials by the following generating function (cf. [2,3,5,6,10,13]):

$$\sum_{n=0}^{\infty} F_n(x,y) \frac{t^n}{n!} = \frac{e^{xt}}{1 - y(e^t - 1)}. \tag{10}$$

When $x = 0$ in (10), the two-variable Fubini polynomials $F_n(x, y)$ reduce to the usual Fubini polynomials given by (cf. [2,3,5,6,10,13]):

$$\sum_{n=0}^{\infty} F_n(y) \frac{t^n}{n!} = \frac{1}{1 - y(e^t - 1)}. \tag{11}$$

It is easy to see that for a non-negative integer n (cf. [2]):

$$F_n\left(x, -\frac{1}{2}\right) = E_n(x), \quad F_n\left(-\frac{1}{2}\right) = E_n \tag{12}$$

and (cf. [3,5,6,10,13]):

$$F_n(y) = \sum_{\mu=0}^{n} S_2(n, \mu) \, \mu! y^{\mu}. \tag{13}$$

Substituting y by 1 in (11), we have the familiar Fubini numbers $F_n(1) := F_n$ as follows (cf. [2,3,5,6,10,13]):

$$\sum_{n=0}^{\infty} F_n \frac{t^n}{n!} = \frac{1}{2 - e^t}. \tag{14}$$

For more information about the applications of the usual Fubini polynomials and numbers, cf. [2,3,5,6,10,13], and see also the references cited therein.

We now define the two-variable truncated Fubini polynomials as follows.

Definition 1. *For non-negative integer m, the two-variable truncated Fubini polynomials are defined via the following exponential generating function:*

$$\sum_{n=0}^{\infty} F_{m,n}(x, y) \frac{t^n}{n!} = \frac{\frac{t^m}{m!} e^{xt}}{1 - y\left(e^t - 1 - \sum_{j=0}^{m-1} \frac{t^j}{j!}\right)}. \tag{15}$$

In the case $x = 0$ in (15), we then get a new type of Fubini polynomial, which we call the truncated Fubini polynomials given by:

$$\sum_{n=0}^{\infty} F_{m,n}(y) \frac{t^n}{n!} = \frac{\frac{t^m}{m!}}{1 - y\left(e^t - 1 - \sum_{j=0}^{m-1} \frac{t^j}{j!}\right)}. \tag{16}$$

Upon setting $x = 0$ and $y = 1$ in (15), we then attain the truncated Fubini numbers $F_{m,n}$ defined by the following Taylor series expansion about $t = 0$:

$$\sum_{n=0}^{\infty} F_{m,n} \frac{t^n}{n!} = \frac{\frac{t^m}{m!}}{2 + \sum_{j=m}^{\infty} \frac{t^j}{j!}}. \tag{17}$$

The two-variable truncated Fubini polynomials $F_{m,n}(x, y)$ cover generalizations of some known polynomials and numbers that we discuss below.

Remark 2. *Setting $m = 0$ in (15), the polynomials $F_{m,n}(x, y)$ reduce to the two-variable Fubini polynomials $F_n(x, y)$ in (10).*

Remark 3. *When $m = 0$ and $x = 0$ in (15), the polynomials $F_{m,n}(x,y)$ become the usual Fubini polynomials $F_n(y)$ in (11).*

Remark 4. *In the special cases $m = 0$, $y = 1$, and $x = 0$ in (15), the polynomials $F_{m,n}(x,y)$ reduce to the familiar Fubini numbers F_n in (14).*

We now are ready to examine the relations and properties for the two-variable Fubini polynomials $F_n(x,y)$, and so, we firstly give the following theorem.

Theorem 1. *The following summation formula:*

$$F_{m,n}(x,y) = \sum_{k=0}^{n} \binom{n}{k} F_{m,k}(y) \, x^{n-k} \tag{18}$$

holds true for non-negative integers m and n.

Proof. By (15), using the Cauchy product in series, we observe that:

$$
\begin{aligned}
\sum_{n=0}^{\infty} F_{m,n}(x,y) \frac{t^n}{n!} &= \frac{\frac{t^m}{m!}}{1 - y\left(e^t - 1 - \sum_{j=0}^{m-1} \frac{t^j}{j!}\right)} e^{xt} \\
&= \sum_{n=0}^{\infty} F_{m,n}(y) \frac{t^n}{n!} \sum_{n=0}^{\infty} x^n \frac{t^n}{n!} \\
&= \sum_{n=0}^{\infty} \sum_{k=0}^{n} \binom{n}{k} F_{m,k}(y) \, x^{n-k} \frac{t^n}{n!},
\end{aligned}
$$

which provides the asserted result (18). $\square$

We now provide another summation formula for the polynomials $F_{m,n}(x,y)$ as follows.

Theorem 2. *The following summation formulas:*

$$F_{m,n}(x+z,y) = \sum_{k=0}^{n} \binom{n}{k} F_{m,k}(x,y) \, z^{n-k} \tag{19}$$

and:

$$F_{m,n}(x+z,y) = \sum_{k=0}^{n} \binom{n}{k} F_{m,k}(y) \, (x+z)^{n-k} \tag{20}$$

are valid for non-negative integers m and n.

Proof. From (15), we obtain:

$$\sum_{n=0}^{\infty} F_{m,n}(x,y)\frac{t^n}{n!} = \frac{\frac{t^m}{m!}}{1-y\left(e^t-1-\sum_{j=0}^{m-1}\frac{t^j}{j!}\right)}e^{(x+z)t}$$

$$= \frac{\frac{t^m}{m!}e^{xt}}{1-y\left(e^t-1-\sum_{j=0}^{m-1}\frac{t^j}{j!}\right)}e^{zt}$$

$$= \sum_{n=0}^{\infty} F_{m,n}(x,y)\frac{t^n}{n!}\sum_{n=0}^{\infty} z^n\frac{t^n}{n!}$$

$$= \sum_{n=0}^{\infty}\sum_{k=0}^{n}\binom{n}{k}F_{m,k}(x,y)z^{n-k}\frac{t^n}{n!}$$

and similarly:

$$\sum_{n=0}^{\infty} F_{m,n}(x,y)\frac{t^n}{n!} = \frac{\frac{t^m}{m!}}{1-y\left(e^t-1-\sum_{j=0}^{m-1}\frac{t^j}{j!}\right)}e^{(x+z)t}$$

$$= \sum_{n=0}^{\infty} F_{m,n}(y)\frac{t^n}{n!}\sum_{n=0}^{\infty}(x+z)^n\frac{t^n}{n!}$$

$$= \sum_{n=0}^{\infty}\sum_{k=0}^{n}\binom{n}{k}F_{m,k}(y)(x+z)^{n-k}\frac{t^n}{n!}$$

which yield the desired results (19) and (20). $\square$

We here define the truncated Stirling numbers of the second kind as follows:

$$\sum_{n=0}^{\infty} S_{2,m}(n,k)\frac{t^n}{n!} = \frac{\left(e^t-1-\sum_{j=0}^{m-1}\frac{t^j}{j!}\right)^k}{k!}. \tag{21}$$

Remark 5. *Upon setting $m=0$ in (21), the truncated Stirling numbers of the second kind $S_{2,m}(n,k)$ reduce to the classical Stirling numbers of the second kind in (8).*

The truncated Stirling numbers of the second kind satisfy the following relationship.

Proposition 1. *The following correlation:*

$$S_{2,m}(n,k+l) = \frac{l!k!}{(k+l)!}\sum_{s=0}^{n}\binom{n}{s}S_{2,m}(s,k)S_{2,m}(n-s,l) \tag{22}$$

holds true for non-negative integers m and n.

Proof. In view of (8) and (21), we have:

$$\sum_{n=0}^{\infty} S_{2,m}\left(n, k+l\right) \frac{t^n}{n!} = \frac{\left(e^t - 1 - \sum_{j=0}^{m-1} \frac{t^j}{j!}\right)^{k+l}}{(k+l)!}$$

$$= \frac{l!\,k!}{(k+l)!} \frac{\left(e^t - 1 - \sum_{j=0}^{m-1} \frac{t^j}{j!}\right)^{k}}{k!} \frac{\left(e^t - 1 - \sum_{j=0}^{m-1} \frac{t^j}{j!}\right)^{l}}{l!}$$

$$= \frac{l!\,k!}{(k+l)!} \sum_{n=0}^{\infty} S_{2,m}\left(n, k\right) \frac{t^n}{n!} \sum_{n=0}^{\infty} S_{2,m}\left(n, l\right) \frac{t^n}{n!}$$

$$= \frac{l!\,k!}{(k+l)!} \sum_{n=0}^{\infty} \sum_{s=0}^{n} \binom{n}{s} S_{2,m}\left(s, k\right) S_{2,m}\left(n-s, l\right) \frac{t^n}{n!},$$

which gives the claimed result (22). $\square$

We present the following correlation between two types of Stirling numbers of the second kind.

Proposition 2. *The following correlation:*

$$S_{2,1}\left(n, k\right) = 2^k S_2\left(n, k : \frac{1}{2}\right) \tag{23}$$

holds true for non-negative integers m and n.

Proof. In view of (8) and (21), we have:

$$\sum_{n=0}^{\infty} S_{2,1}\left(n, k\right) \frac{t^n}{n!} = \frac{\left(e^t - 1 - 1\right)^{k}}{k!}$$

$$= \frac{2^k \left(\frac{1}{2} e^t - 1\right)^{k}}{k!}$$

$$= 2^k \sum_{n=0}^{\infty} S_2\left(n, k : \frac{1}{2}\right) \frac{t^n}{n!},$$

which presents the desired result (23). $\square$

A relation that includes $F_{m,n}\left(x\right)$ and $S_{2,m}\left(n, k\right)$ is given by the following theorem.

Theorem 3. *The following relation:*

$$F_{m,n+m}\left(x\right) = \sum_{k=0}^{n} \binom{n+m}{m} x^k k! S_{2,m}\left(n, k\right) \tag{24}$$

is valid for a complex number x with $|x| < 1$ and non-negative integers m and n.

Proof. By (16) and (21), we see that:

$$\sum_{n=0}^{\infty} F_{m,n}(x) \frac{t^n}{n!} = \frac{\frac{t^m}{m!}}{1 - x\left(e^t - 1 - \sum_{j=0}^{m-1} \frac{t^j}{j!}\right)}$$

$$= \frac{t^m}{m!} \sum_{k=0}^{\infty} x^k \left(e^t - 1 - \sum_{j=0}^{m-1} \frac{t^j}{j!}\right)^k$$

$$= \frac{t^m}{m!} \sum_{k=0}^{\infty} x^k k! \sum_{n=0}^{\infty} S_{2,m}(n,k) \frac{t^n}{n!}$$

$$= \sum_{n=0}^{\infty} \sum_{k=0}^{\infty} x^k k! S_{2,m}(n,k) \frac{t^{n+m}}{m!n!},$$

which implies the desired result (24). $\square$

We now state the following theorem.

Theorem 4. *The following identity:*

$$F_{1,n+1}(x) = n \sum_{k=1}^{\infty} x^k k! S_2\left(n, k : \frac{1}{2}\right) \tag{25}$$

holds true for a complex number x with $|x| < 1$ and a positive integer n.

Proof. By (9) an (16), using the Cauchy product in series, we observe that:

$$\sum_{n=0}^{\infty} F_{1,n}(x) \frac{t^n}{n!} = \frac{t}{1 - x(e^t - 2)}$$

$$= t \sum_{k=0}^{\infty} x^k (e^t - 2)^k$$

$$= t \sum_{k=0}^{\infty} x^k k! \frac{\left(\frac{1}{2}e^t - 1\right)^k}{k!} 2^k$$

$$= \sum_{n=0}^{\infty} \sum_{k=0}^{\infty} x^k k! S_2\left(n, k : \frac{1}{2}\right) \frac{t^{n+1}}{n!},$$

which provides the asserted result (25). $\square$

We now provide the derivative property for the polynomials $F_{m,n}(x,y)$ as follows.

Theorem 5. *The derivative formula:*

$$\frac{\partial}{\partial x} F_{m,n}(x,y) = n F_{m,n-1}(x,y) \tag{26}$$

holds true for non-negative integers m and a positive integer n.

Proof. Applying the derivative operator with respect to x to both sides of the equation (15), we acquire:

$$\frac{\partial}{\partial x}\left(\sum_{n=0}^{\infty} F_{m,n}(x,y)\frac{t^n}{n!}\right) = \frac{\partial}{\partial x}\left(\frac{\frac{t^m}{m!}e^{xt}}{1-y\left(e^t-1-\sum_{j=0}^{m-1}\frac{t^j}{j!}\right)}\right)$$

and then:

$$
\begin{aligned}
\sum_{n=0}^{\infty}\frac{\partial}{\partial x}F_{m,n}(x,y)\frac{t^n}{n!} &= \frac{\frac{t^m}{m!}}{1-y\left(e^t-1-\sum_{j=0}^{m-1}\frac{t^j}{j!}\right)}\frac{\partial}{\partial x}e^{xt}\\
&= \frac{\frac{t^m}{m!}e^{xt}}{1-y\left(e^t-1-\sum_{j=0}^{m-1}\frac{t^j}{j!}\right)}t\\
&= \sum_{n=0}^{\infty}F_{m,n}(x,y)\frac{t^{n+1}}{n!},
\end{aligned}
$$

which means the claimed result (26). $\square$

A recurrence relation for the two-variable truncated Fubini polynomials is given by the following theorem.

Theorem 6. *The following equalities:*

$$F_{m,n}(x,y) = 0 \quad (n = 0,1,2,\cdots,m-1)$$

and:

$$F_{m,n+m}(x,y) = \frac{y}{1+y}\sum_{s=0}^{n}\binom{n+m}{s}F_{m,s}(x,y) - \frac{x^n}{1+y}\frac{(n+m)!}{n!m!} \tag{27}$$

hold true for non-negative integers m and n.

Proof. Using Definition 1, we can write:

$$
\begin{aligned}
\frac{t^m}{m!}e^{xt} &= \left(1-y\left(\sum_{j=m}^{\infty}\frac{t^j}{j!}-1\right)\right)\sum_{n=0}^{\infty}F_{m,n}(x,y)\frac{t^n}{n!}\\
&= \sum_{n=0}^{\infty}F_{m,n}(x,y)\frac{t^n}{n!} - y\left[\sum_{j=m}^{\infty}\frac{t^j}{j!}\sum_{n=0}^{\infty}F_{m,n}(x,y)\frac{t^n}{n!} - \sum_{n=0}^{\infty}F_{m,n}(x,y)\frac{t^n}{n!}\right]\\
&= \sum_{n=0}^{\infty}F_{m,n}(x,y)\frac{t^n}{n!} - y\left[\sum_{j=0}^{\infty}\frac{t^{j+m}}{(j+m)!}\sum_{n=0}^{\infty}F_{m,n}(x,y)\frac{t^n}{n!} - \sum_{n=0}^{\infty}F_{m,n}(x,y)\frac{t^n}{n!}\right].
\end{aligned}
$$

Because of:

$$\sum_{j=0}^{\infty}\frac{t^{j+m}}{(j+m)!}\sum_{n=0}^{\infty}F_{m,n}(x,y)\frac{t^n}{n!} = \sum_{n=0}^{\infty}\sum_{j=0}^{n}\binom{n+m}{j}F_{m,j}(x,y)\frac{t^{n+m}}{(n+m)!},$$

we obtain:

$$\sum_{n=0}^{\infty} x^n \frac{t^{n+m}}{n!\,m!} = \sum_{n=0}^{\infty} F_{m,n}\,(x,y)\,\frac{t^n}{n!} - y \sum_{n=0}^{\infty} \sum_{j=0}^{n} \binom{n+m}{j} F_{m,j}\,(x,y)\,\frac{t^{n+m}}{(n+m)!}$$

$$+ y \sum_{n=0}^{\infty} F_{m,n}\,(x,y)\,\frac{t^n}{n!}.$$

Thus, we arrive at the following equality:

$$\sum_{n=0}^{\infty} F_{m,n}\,(x,y)\,\frac{t^n}{n!} = \frac{1}{1+y} \sum_{n=0}^{\infty} \left(y \sum_{j=0}^{n} \binom{n+m}{j} \frac{F_{m,j}\,(x,y)}{(n+m)!} - \frac{x^n}{n!\,m!} \right) t^{n+m}.$$

Comparing the coefficients of both sides of the last equality, the proof is completed. $\square$

Theorem 6 can be used to determine the two-variable truncated Fubini polynomials. Thus, we provide some examples as follows.

Example 1. *Choosing $m = 1$, then we have $F_{1,0}\,(x,y) = 0$. Utilizing the recurrence formula (27), we derive:*

$$F_{1,n+1}\,(x,y) = \frac{y}{1-y} \sum_{s=0}^{n} \binom{n+1}{s} F_{1,s}\,(x,y) - \frac{x^n}{1-y}\,(n+1).$$

Thus, we subsequently acquire:

$$F_{1,1}\,(x,y) = -\frac{1}{1+y},$$

$$F_{1,2}\,(x,y) = -\frac{2y}{(1+y)^2} - \frac{2x}{1+y},$$

$$F_{1,3}\,(x,y) = \frac{3}{1+y} \left(\frac{2y^2}{(1+y)^2} - \frac{2xy}{1+y} - x^2 \right).$$

Furthermore, choosing $m = 2$, we then obtain the following recurrence relation:

$$F_{2,n+2}\,(x,y) = \frac{y}{1+y} \sum_{s=0}^{n} \binom{n+2}{s} F_{2,s}\,(x,y) - \frac{x^n}{1+y}\,\frac{(n+2)\,(n+1)}{2}$$

which yields the following polynomials:

$$F_{2,0}\,(x,y) = F_{2,1}\,(x,y) = 0,$$

$$F_{2,2}\,(x,y) = -\frac{1}{1+y},$$

$$F_{2,3}\,(x,y) = -\frac{3x}{1+y},$$

$$F_{2,4}\,(x,y) = \frac{6x}{y+1} \left(\frac{3y}{1+y} + x \right).$$

By applying a similar method used above, one can derive the other two-variable truncated Fubini polynomials.

Here is a correlation that includes the truncated Fubini polynomials and Stirling numbers of the second kind.

Theorem 7. *For non-negative integers n and m, we have:*

$$F_{m,n}(x,y) = \sum_{u=0}^{n} \sum_{k=0}^{u} \binom{n}{u} F_{m,n-u}(y) \, S_2(u,k) \, (x)_k. \tag{28}$$

Proof. By means of Theorem 1 and Formula (8), we get:

$$
\begin{aligned}
F_{m,n}(x,y) &= \sum_{u=0}^{n} \binom{n}{u} F_{m,n-u}(y) \, x^u \\
&= \sum_{u=0}^{n} \binom{n}{u} F_{m,n-u}(y) \sum_{k=0}^{u} S_2(u,k) \, (x)_k,
\end{aligned}
$$

which completes the proof of this theorem. $\square$

The rising factorial number x is defined by $(x)^{(n)} = x(x+1)(x+2)\cdots(x+n-1)$ for a positive integer n. We also note that the negative binomial expansion is given as follows:

$$(x+a)^{-n} = \sum_{k=0}^{\infty} (-1)^k \binom{n+k-1}{k} x^k a^{-n-k} \tag{29}$$

for negative integer $-n$ and $|x| < a$; cf. [7].

Here, we give the following theorem.

Theorem 8. *The following relationship:*

$$F_{m,n}(x,y) = \sum_{k=0}^{\infty} \sum_{l=k}^{n} \binom{n}{l} S_2(l,k) \, F_{n,n-l}(-k,y) \, (x)^{(k)} \tag{30}$$

holds true for non-negative integers n and m.

Proof. By means of Definition 1 and using Equations (7) and (29), we attain:

$$
\begin{aligned}
\sum_{n=0}^{\infty} F_{m,n}(x,y) \frac{t^n}{n!}
&= \frac{\frac{t^m}{m!}}{1 - y\left(e^t - 1 - \sum_{j=0}^{m-1} \frac{t^j}{j!}\right)} \left(e^{-t}\right)^{-x} \\
&= \frac{\frac{t^m}{m!}}{1 - y\left(e^t - 1 - \sum_{j=0}^{m-1} \frac{t^j}{j!}\right)} \sum_{k=0}^{\infty} \binom{x+k-1}{k} \left(1 - e^{-t}\right)^{-k} \\
&= \frac{\frac{t^m}{m!}}{1 - y\left(e^t - 1 - \sum_{j=0}^{m-1} \frac{t^j}{j!}\right)} \sum_{k=0}^{\infty} (x)^{(k)} \frac{\left(e^t - 1\right)^k}{k!} e^{-kt} \\
&= \sum_{k=0}^{\infty} (x)^{(k)} \sum_{n=0}^{\infty} F_{m,n}(-k,y) \frac{t^n}{n!} \sum_{n=0}^{\infty} S_2(n,k) \frac{t^n}{n!} \\
&= \sum_{k=0}^{\infty} (x)^{(k)} \sum_{n=0}^{\infty} \left(\sum_{l=0}^{n} \binom{n}{l} F_{m,n-l}(-k,y) \, S_2(l,k)\right) \frac{t^n}{n!},
\end{aligned}
$$

which gives the asserted result (30). $\square$

Therefore, we give the following theorem.

Theorem 9. *The following relationship:*

$$y \sum_{k=0}^{n} \binom{n}{k} F_{m,n-k}(z,y) F_{m+1,k}(x,y) = \sum_{k=0}^{n} \binom{n}{k} F_{m+1,n-k}(x,y) z^k \tag{31}$$

$$-\frac{n}{m+1} \sum_{k=0}^{n-1} \binom{n-1}{k} F_{m,n-1-k}(z,y) x^k$$

holds true for non-negative integers n and m.

Proof. By means of Definition 1, we see that:

$$e^{xt} \frac{t^{m+1}}{(m+1)!} = \left(1 - y\left(e^t - 1 - \sum_{j=0}^{m} \frac{t^j}{j!}\right)\right) \sum_{n=0}^{\infty} F_{m+1,n}(x,y) \frac{t^n}{n!}$$

$$= \left(1 - y\left(e^t - 1 - \sum_{j=0}^{m-1} \frac{t^j}{j!}\right)\right) \sum_{n=0}^{\infty} F_{m+1,n}(x,y) \frac{t^n}{n!}$$

$$-y \frac{t^m}{m!} \sum_{n=0}^{\infty} F_{m+1,n}(x,y) \frac{t^n}{n!}.$$

Thus, we get:

$$e^{xt} \frac{t^{m+1}}{(m+1)!} \sum_{n=0}^{\infty} F_{m,n}(z,y) \frac{t^n}{n!} = \frac{t^m}{m!} e^{zt} \sum_{n=0}^{\infty} F_{m+1,n}(x,y) \frac{t^n}{n!}$$

$$-y \frac{t^m}{m!} \sum_{n=0}^{\infty} F_{m,n}(z,y) \frac{t^n}{n!} \sum_{n=0}^{\infty} F_{m+1,n}(x,y) \frac{t^n}{n!}$$

and then:

$$\sum_{n=0}^{\infty} \sum_{k=0}^{n} \binom{n}{k} F_{m,n-k}(z,y) x^k \frac{t^{n+1}}{n!(m+1)} = \sum_{n=0}^{\infty} \sum_{k=0}^{n} \binom{n}{k} F_{m+1,n-k}(x,y) z^k \frac{t^n}{n!}$$

$$-y \sum_{n=0}^{\infty} y \sum_{k=0}^{n} \binom{n}{k} F_{m,n-k}(z,y) F_{m+1,k}(x,y) \frac{t^n}{n!}$$

which provides the claimed result in (31). $\square$

Here, we investigate a linear combination for the two-variable truncated Fubini polynomials for different y values in the following theorem.

Theorem 10. *Let the numbers m and n be non-negative integers and $y_1 \neq y_2$. We then have:*

$$\frac{m!n!}{(n+m)!} \sum_{k=0}^{n+m} \binom{n+m}{k} F_{m,n+m-k}(x_1,y_1) F_{m,k}(x_2,y_2) = \frac{y_2 F_{m,n-k}(x_1+x_2,y_2) - y_1 F_{m,n-k}(x_1+x_2,y_1)}{y_2 - y_1}. \tag{32}$$

Proof. By Definition 1, we consider the following product:

$$
\frac{\frac{t^m}{m!}e^{x_1 t}}{1-y_1\left(e^t-1-\sum_{j=0}^{m-1}\frac{t^j}{j!}\right)}\frac{\frac{t^m}{m!}e^{x_2 t}}{1-y_2\left(e^t-1-\sum_{j=0}^{m-1}\frac{t^j}{j!}\right)}
$$

$$
=\frac{y_2}{y_2-y_1}\frac{\frac{t^{2m}}{(m!)^2}e^{(x_1+x_2)t}}{1-y_2\left(e^t-1-\sum_{j=0}^{m-1}\frac{t^j}{j!}\right)}-\frac{y_1}{y_2-y_1}\frac{\frac{t^{2m}}{(m!)^2}e^{(x_1+x_2)t}}{1-y_1\left(e^t-1-\sum_{j=0}^{m-1}\frac{t^j}{j!}\right)},
$$

which yields

$$
\sum_{n=0}^{\infty}\sum_{k=0}^{n}\binom{n}{k}F_{m,n-k}\left(x_1,y_1\right)F_{m,k}\left(x_2,y_2\right)\frac{t^n}{n!}
$$

$$
=\frac{y_2}{y_2-y_1}\sum_{n=0}^{\infty}F_{m,n}\left(x_1+x_2,y_2\right)\frac{t^{n+m}}{n!m!}-\frac{y_1}{y_2-y_1}\sum_{n=0}^{\infty}F_{m,n}\left(x_1+x_2,y_1\right)\frac{t^{n+m}}{n!m!}.
$$

Thus, we get:

$$
\sum_{n=0}^{\infty}\left(\sum_{k=0}^{n}\binom{n}{k}F_{m,n-k}\left(x_1,y_1\right)F_{m,k}\left(x_2,y_2\right)\right)\frac{t^n}{n!}
$$

$$
=\sum_{n=0}^{\infty}\left(\frac{y_2}{y_2-y_1}F_{m,n}\left(x_1+x_2,y_2\right)-\frac{y_1}{y_2-y_1}F_{m,n}\left(x_1+x_2,y_1\right)\right)\frac{t^{n+m}}{n!m!},
$$

which gives the desired result (32). $\quad\square$

3. Correlations with Truncated Euler and Bernoulli Polynomials

In this section, we investigate several correlations for the two-variable truncated Fubini polynomials $F_{m,n}\left(x,y\right)$ related to the truncated Euler polynomials $E_{m,n}\left(x\right)$ and numbers $E_{m,n}$ and the truncated Bernoulli polynomials $B_{m,n}\left(x\right)$ and numbers $B_{m,n}$.

Here is a relation between the truncated Euler polynomials and two-variable truncated Fubini polynomials at the special value $y=-\frac{1}{2}$.

Theorem 11. *We have:*

$$
F_{m,n}\left(x,-\frac{1}{2}\right)=E_{m,n}\left(x\right). \tag{33}
$$

Proof. In terms of (5) and (15), we get:

$$
\sum_{n=0}^{\infty}F_{m,n}\left(x,-\frac{1}{2}\right)\frac{t^n}{n!}=\frac{\frac{t^m}{m!}e^{xt}}{1+\frac{1}{2}\left(e^t-1-\sum_{j=0}^{m-1}\frac{t^j}{j!}\right)}
$$

$$
=\frac{2\frac{t^m}{m!}e^{xt}}{e^t+1-\sum_{j=0}^{m-1}\frac{t^j}{j!}}
$$

$$
=\sum_{n=0}^{\infty}E_{m,n}\left(x\right)\frac{t^n}{n!},
$$

which implies the asserted result (33). $\quad\square$

Corollary 1. *Taking $x = 0$, we then get a relation between the truncated Euler numbers and truncated Fubini polynomials at the special value $y = -\frac{1}{2}$, namely:*

$$F_{m,n}\left(-\frac{1}{2}\right) = E_{m,n}. \tag{34}$$

Remark 6. *The relations (33) and (34) are extensions of the relations in (12).*

We now state the following theorem, which includes a correlation for $F_{m,n}(x,y)$, $F_{m,n}(y)$ and $E_{m,n}(x)$.

Theorem 12. *The following formula:*

$$
\begin{aligned}
F_{m,n}(x,y) \;=\;& \frac{n!\,m!}{(n+m)!} \sum_{l=0}^{n+m} \frac{1}{2}\binom{n+m}{l} F_{m,l}(y)\, E_{m,n+m-l}(x) \\[4pt]
&+ \frac{n!\,m!}{(n+m)!} \sum_{j=0}^{n} \frac{1}{2}\binom{n+m}{j} \sum_{l=0}^{j}\binom{j}{l} F_{m,l}(y)\, E_{m,j-l}(x)
\end{aligned}
\tag{35}
$$

is valid for non-negative integers m and n.

Proof. By (5) and (15), we acquire that:

$$
\begin{aligned}
\sum_{n=0}^{\infty} F_{m,n}(x,y)\,\frac{t^n}{n!} \;=\;& \frac{\frac{t^m}{m!}e^{xt}}{1 - y\left(e^t - 1 - \sum_{j=0}^{m-1}\frac{t^j}{j!}\right)}\cdot\frac{2\frac{t^m}{m!}}{e^t + 1 - \sum_{j=0}^{m-1}\frac{t^j}{j!}}\cdot\frac{e^t + 1 - \sum_{j=0}^{m-1}\frac{t^j}{j!}}{2\frac{t^m}{m!}} \\[6pt]
=\;& \frac{1}{2}\frac{m!}{t^m}\sum_{n=0}^{\infty} F_{m,n}(y)\,\frac{t^n}{n!}\sum_{n=0}^{\infty} E_{m,n}(x)\,\frac{t^n}{n!}\left(\sum_{j=m}^{\infty}\frac{t^j}{j!}+1\right) \\[6pt]
=\;& \frac{m!}{2}\sum_{n=0}^{\infty}\left(\sum_{l=0}^{n}\binom{n}{l} F_{m,l}(y)\, E_{m,n-l}(x)\right)\frac{t^{n-m}}{n!}\left(\sum_{j=0}^{\infty}\frac{t^{j+m}}{(j+m)!}+1\right) \\[6pt]
=\;& \frac{m!}{2}\sum_{n=0}^{\infty}\left(\sum_{l=0}^{n}\binom{n}{l} F_{m,l}(y)\, E_{m,n-l}(x)\right)\frac{t^{n-m}}{n!} \\[6pt]
&+ \frac{m!}{2}\sum_{n=0}^{\infty}\sum_{j=0}^{n}\binom{n+m}{j}\left(\sum_{l=0}^{j}\binom{j}{l} F_{m,l}(y)\, E_{m,j-l}(x)\right)\frac{t^n}{(n+m)!},
\end{aligned}
$$

which completes the proof of the theorem. $\square$

We finally state the relations for the truncated Bernoulli and Fubini polynomials as follows.

Theorem 13. *The following relation:*

$$F_{m,n}(x,y) = \frac{n!\,m!}{(n+m)!}\sum_{l=0}^{n}\binom{n+m}{l}\sum_{k=0}^{l}\binom{l}{k} F_{m,l}(y)\, B_{m,l-k}(x) \tag{36}$$

is valid for non-negative integers m and n.

Proof. By (5) and (15), we acquire that:

$$
\begin{aligned}
\sum_{n=0}^{\infty} F_{m,n}(x,y)\frac{t^n}{n!} &= \frac{\frac{t^m}{m!}e^{xt}}{1-y\left(e^t-1-\sum_{j=0}^{m-1}\frac{t^j}{j!}\right)e^t-\sum_{j=0}^{m-1}\frac{t^j}{j!}}\frac{\frac{t^m}{m!}}{e^t-\sum_{j=0}^{m-1}\frac{t^j}{j!}}\frac{e^t-\sum_{j=0}^{m-1}\frac{t^j}{j!}}{\frac{t^m}{m!}} \\
&= \frac{m!}{t^m}\sum_{n=0}^{\infty}F_{m,n}(y)\frac{t^n}{n!}\sum_{n=0}^{\infty}B_{m,n}(x)\frac{t^n}{n!}\sum_{j=m}^{\infty}\frac{t^j}{j!} \\
&= m!\sum_{n=0}^{\infty}\left(\sum_{k=0}^{n}\binom{n}{k}F_{m,k}(y)B_{m,n-k}(x)\right)\frac{t^n}{n!}\sum_{j=0}^{\infty}\frac{t^j}{(j+m)!} \\
&= \frac{n!\,m!}{(n+m)!}\sum_{n=0}^{\infty}\sum_{l=0}^{n}\binom{n+m}{l}\left(\sum_{k=0}^{l}\binom{l}{k}F_{m,l}(y)B_{m,l-k}(x)\right)\frac{t^n}{(n+m)!},
\end{aligned}
$$

which means the asserted result (36). $\square$

4. Conclusions

In this paper, we firstly considered two-variable truncated Fubini polynomials and numbers, and we then obtained some identities and properties for these polynomials and numbers, involving summation formulas, recurrence relations, and the derivative property. We also proved some formulas related to the truncated Stirling numbers of the second kind and Apostol-type Stirling numbers of the second kind. Furthermore, we gave some correlations including the two-variable truncated Fubini polynomials, the truncated Euler polynomials, and truncated Bernoulli polynomials.

Author Contributions: Both authors have equally contributed to this work. Both authors read and approved the final manuscript.

Funding: This research received no external funding.

Conflicts of Interest: The authors declare no conflict of interest.

References

1. Hassen, A.; Nguyen, H.D. Hypergeometric Bernoulli polynomials and Appell sequences. *Int. J. Number Theory* **2008**, *4*, 767–774. [CrossRef]
2. Kargın, L. Exponential polynomials and its applications to the related polynomials and numbers. *arXiv* **2016**, arXiv:1503.05444v2.
3. Kargın, L. Some formulae for products of Fubini polynomials with applications. *arXiv* **2016**, arXiv:1701.01023v1.
4. Khan, S.; Yasmin, G.; Ahmad, N. A note on truncated exponential-based Appell polynomials. *Bull. Malays. Math. Sci. Soc.* **2017**, *40*, 373–388. [CrossRef]
5. Kilar, N.; Simsek, Y. A new family of Fubini type numbers and polynomials associated with Apostol-Bernoulli numbers and polynomials. *J. Korean Math. Soc.* **2017**, *54*, 1605–1621.
6. Kim, D.S.; Kim, T.; Kwon, H.-I.; Park, J.-W. Two variable higher-order Fubini polynomials. *J. Korean Math. Soc.* **2018**, *55*, 975–986.
7. Komatsu, T.; Ruiz, C.D.J.P. Truncated Euler polynomials. *Math. Slovaca* **2018**, *68*, 527–536. [CrossRef]
8. Luo, Q.-M.; Srivastava, H.M. Some generalizations of the Apostol-Genocchi polynomials and the Stirling numbers of the second kind. *Appl. Math. Comput.* **2011**, *217*, 5702–5728. [CrossRef]
9. Srivastava, H.M.; Araci, S.; Khan, W.A.; Acikgoz, M. A note on the truncated-exponential based Apostol-type polynomials. *Symmetry* **2019**, *11*, 538. [CrossRef]
10. Su, D.D.; He, Y. Some identities for the two variable Fubini polynomials. *Mathematics* **2019**, *7*, 115. [CrossRef]

11. Dattoli, G.; Ceserano, C.; Sacchetti, D. A note on truncated polynomials. *Appl. Math. Comput.* **2003**, *134*, 595–605. [CrossRef]
12. Yasmin, G.; Khan, S.; Ahmad, N. Operational methods and truncated exponential-based Mittag-Leffler polynomials. *Mediterr. J. Math.* **2016**, *13*, 1555. [CrossRef]
13. Kim, T.; Kim, D.S.; Jang, G.-W. A note on degenerate Fubini polynomials. *Proc. Jangjeon Math. Soc.* **2017**, *20*, 521–531.

Article

On Positive Quadratic Hyponormality of a Unilateral Weighted Shift with Recursively Generated by Five Weights

Chunji Li [1] and Cheon Seoung Ryoo [2],*

[1] Department of Mathematics, Northeastern University, Shenyang 110-004, China; lichunji@mail.neu.edu.cn
[2] Department of Mathematics, Hannam University, Daejeon 34430, Korea
* Correspondence: ryoocs@hnu.kr

Received: 16 January 2019; Accepted: 8 February 2019; Published: 25 February 2019

Abstract: Let $1 < a < b < c < d$ and $\hat{\alpha}_{[5]} := \left(1, \sqrt{a}, \sqrt{b}, \sqrt{c}, \sqrt{d}\right)^{\wedge}$ be a weighted sequence that is recursively generated by five weights $1, \sqrt{a}, \sqrt{b}, \sqrt{c}, \sqrt{d}$. In this paper, we give sufficient conditions for the positive quadratic hyponormalities of $W_{\alpha(x)}$ and $W_{\alpha(y,x)}$, with $\alpha(x) : \sqrt{x}, \hat{\alpha}_{[5]}$ and $\alpha(y,x) : \sqrt{y}, \sqrt{x}, \hat{\alpha}_{[5]}$.

Keywords: positively quadratically hyponormal; quadratically hyponormal; unilateral weighted shift; recursively generated

MSC: 47B37; 47B20

---

1. Introduction

Let $\mathcal{H}$ be a separable, infinite dimensional, complex Hilbert space, and let $\mathcal{L}(\mathcal{H})$ be the algebra of all bounded linear operators on $\mathcal{H}$. An operator T in $\mathcal{L}(\mathcal{H})$ is said to be *normal* if $T^*T = TT^*$, *hyponormal* if $T^*T \geq TT^*$, and *subnormal* if $T = N|_{\mathcal{H}}$, where N is normal on some Hilbert space $K \supseteq \mathcal{H}$. For $A, B \in \mathcal{L}(\mathcal{H})$, let $[A, B] := AB - BA$. We say that an n-tuple $T = (T_1, \ldots, T_n)$ of operators in $\mathcal{L}(\mathcal{H})$ is *hyponormal* if the operator matrix $([T_j^*, T_i])_{i,j=1}^n$ is positive on the direct sum of n copies of $\mathcal{H}$. For arbitrary positive integer k, $T \in \mathcal{L}(\mathcal{H})$ is (strongly) *k-hyponormal* if $(I, T, \ldots, T^k)$ is hyponormal. It is well known that T is subnormal if and only if T is ∞-hyponormal. An operator T in $\mathcal{L}(\mathcal{H})$ is said to be *weakly n-hyponormal* if $p(T)$ is hyponormal for any polynomial p with degree less than or equal to n. An operator T is *polynomially hyponormal* if $p(T)$ is hyponormal for every polynomial p. In particular, the weak two-hyponormality (or weak three-hyponormality) is referred to as quadratical hyponormality (or cubical hyponormality, resp.) and has been considered in detail in [1–9].

Let $\{e_n\}_{n=0}^{\infty}$ be the canonical orthonormal basis for Hilbert space $l^2(\mathbb{Z}_+)$, and let $\alpha := \{\alpha_n\}_{n=0}^{\infty}$ be a bounded sequence of positive numbers. Let W_α be a unilateral weighted shift defined by $W_\alpha e_n := \alpha_n e_{n+1} \, (n \geq 0)$. It is well known that W_α is hyponormal if and only if $\alpha_n \leq \alpha_{n+1} \, (n \geq 0)$. The moments of W_α are usually defined by $\gamma_0 := 1, \gamma_i := \alpha_0^2 \cdots \alpha_{i-1}^2 \, (i \geq 1)$. It is well known that W_α is subnormal if and only if there exists a Borel probability measure μ supported in $\left[0, \|W_\alpha\|^2\right]$, with $\|W_\alpha\|^2 \in \mathrm{supp}\, \mu$, such that [10] $\gamma_n = \int t^n d\mu(t)\ (\forall n \geq 0)$. It follows from [11] (Theorem 4) that W_α is subnormal if and only if for every $k \geq 1$ and every $n \geq 0$, the Hankel matrix:

$$A(n,k) := \begin{bmatrix} \gamma_n & \gamma_{n+1} & \gamma_{n+2} & \cdots & \gamma_{n+k} \\ \gamma_{n+1} & \gamma_{n+2} & \gamma_{n+3} & \cdots & \gamma_{n+k+1} \\ \gamma_{n+2} & \gamma_{n+3} & \gamma_{n+4} & \cdots & \gamma_{n+k+2} \\ \vdots & \vdots & \vdots & \ddots & \vdots \\ \gamma_{n+k} & \gamma_{n+k+1} & \gamma_{n+k+2} & \cdots & \gamma_{n+2k} \end{bmatrix} \geq 0.$$

A weighted shift W_α is said to be *recursively generated* if there exists $i \geq 1$ and $\varphi = (\varphi_0, \ldots, \varphi_{i-1}) \in \mathbb{C}^i$ such that:

$$\gamma_n = \varphi_{i-1}\gamma_{n-1} + \cdots + \varphi_0 \gamma_{n-i} \quad (n \geq i),$$

where γ_n is the moment of W_α, i.e., $\gamma_0 := 1, \gamma_i := \alpha_0^2 \cdots \alpha_{i-1}^2\ (i \geq 1)$, equivalently,

$$\alpha_n^2 = \varphi_{i-1} + \frac{\varphi_{i-2}}{\alpha_{n-1}^2} + \cdots + \frac{\varphi_0}{\alpha_{n-1}^2 \cdots \alpha_{n-i+1}^2} \quad (n \geq i).$$

Given an initial segment of weights $\alpha : \alpha_0, \ldots, \alpha_{2k}\ (k \geq 0)$, there is a canonical procedure to generate a sequence (denoted $\hat{\alpha}$) in such a way that $W_{\hat{\alpha}}$ is a recursively-generated shift having α as an initial segment of weight. In particular, given an initial segment of weights $\alpha : \sqrt{a}, \sqrt{b}, \sqrt{c}$ with $0 < a < b < c$, we obtain $\varphi_0 = -\frac{ab(c-b)}{b-a}$ and $\varphi_1 = \frac{b(c-a)}{b-a}$.

In [12,13], Curto-Putinar proved that there exists an operator that is polynomially hyponormal, but not two-hyponormal. Although the existence of a weighted shift, which is polynomially hyponormal, but not subnormal, was established in [12,13], a concrete example of such weighted shifts has not been found yet. Recently, the authors in [14] proved that the subnormality is equivalent to the polynomial hyponormality for recursively-weighted shift W_α with $\alpha : \sqrt{x}, \left(\sqrt{a}, \sqrt{b}, \sqrt{c}\right)^\wedge$. Based on this, in this paper, we have to consider the weighted shift operator with five generated elements.

The organization of this paper is as follows. In Section 2, we recall some terminology and notations concerning the quadratic hyponormality and positive quadratic hyponormality of unilateral weighted shifts W_α. In Section 3, we give some results on the unilateral weighted shifts with recursively generated by five weights $\hat{\alpha}_{[5]} := \left(1, \sqrt{a}, \sqrt{b}, \sqrt{c}, \sqrt{d}\right)^\wedge\ (1 < a < b < c < d)$. In Section 4, we consider positive quadratic hyponormalities of W_α with weights $\alpha : \sqrt{x}, \hat{\alpha}_{[5]}$ and $\alpha : \sqrt{y}, \sqrt{x}, \hat{\alpha}_{[5]}$. In Section 5, we give more results on the positive quadratic hyponormality for any unilateral weighted shift W_α. In Section 6, we present the conclusions.

2. Preliminaries and Notations

Recall that a weighted shift W_α is *quadratically hyponormal* if $W_\alpha + sW_\alpha^2$ is hyponormal for any $s \in \mathbb{C}$ [2], i.e., $D(s) := [(W_\alpha + sW_\alpha^2)^*, W_\alpha + sW_\alpha^2] \geq 0$, for any $s \in \mathbb{C}$. Let $\{e_i\}_{i=0}^\infty$ be an orthonormal basis for $\mathcal{H}$, and let P_n be the orthogonal projection on $\vee_{i=0}^n \{e_i\}$. For $s \in \mathbb{C}$, we let:

$$\begin{aligned} D_n(s) &= P_n[(W_\alpha + sW_\alpha^2)^*, W_\alpha + sW_\alpha^2]P_n \\ &= \begin{bmatrix} q_0 & r_0 & 0 & \cdots & 0 & 0 \\ \bar{r}_0 & q_1 & r_1 & \cdots & 0 & 0 \\ 0 & \bar{r}_1 & q_2 & \ddots & 0 & 0 \\ \vdots & \vdots & \ddots & \ddots & \ddots & \vdots \\ 0 & 0 & 0 & \ddots & q_{n-1} & r_{n-1} \\ 0 & 0 & 0 & \cdots & \bar{r}_{n-1} & q_n \end{bmatrix}, \end{aligned}$$

where:

$$\begin{aligned}
q_k &: \; = u_k + |s|^2 v_k, \quad r_k := w_k \bar{s}, \\
u_k &: \; = \alpha_k^2 - \alpha_{k-1}^2, \quad v_k := \alpha_k^2 \alpha_{k+1}^2 - \alpha_{k-2}^2 \alpha_{k-1}^2, \\
w_k &: \; = \alpha_k^2 (\alpha_{k+1}^2 - \alpha_{k-1}^2)^2, \quad \text{for } k \geq 0,
\end{aligned}$$

and $\alpha_{-1} = \alpha_{-2} := 0$. Hence, W_α is quadratically hyponormal if and only if $D_n(s) \geq 0$ for every $s \in \mathbb{C}$ and every $n \geq 0$. Hence, we consider $d_n(\cdot) := \det D_n(\cdot)$, which is a polynomial in $t := |s|^2$ of degree $n + 1$, with Maclaurin expansion $d_n(t) := \sum_{i=0}^{n+1} c(n,i) t^i$. It is easy to find the following recursive relations [2]:

$$\begin{cases}
d_0(t) = q_0, \\
d_1(t) = q_0 q_1 - |r_0|^2, \\
d_{n+2}(t) = q_{n+2} d_{n+1}(t) - |r_{n+1}|^2 d_n(t) \quad (n \geq 0).
\end{cases}$$

Furthermore, we can obtain the following:

$$\begin{aligned}
c(0,0) &= u_0, \; c(0,1) = v_0, \\
c(1,0) &= u_1 u_0, \; c(1,1) = u_1 v_0 + u_0 v_1 - w_0, \; c(1,2) = v_1 v_0,
\end{aligned}$$

and:

$$\begin{aligned}
c(n+2,i) &= u_{n+2} c(n+1,i) + v_{n+2} c(n+1,i-1) - w_{n+1} c(n,i-1) \\
&\quad (n \geq 0, \text{ and } 0 \leq i \leq n+1).
\end{aligned}$$

In particular, for any $n \geq 0$, we have:

$$c(n,0) = u_0 u_1 \cdots u_n, \; c(n,n+1) = v_0 v_1 \cdots v_n.$$

Furthermore, we can obtain the following results.

Lemma 1. *Let $\rho := v_2(u_0 v_1 - w_0) + v_0(u_1 v_2 - w_1)$. Then, for any $n \geq 4$, we have:*

$$\begin{aligned}
c(n,n) &= u_n c(n-1,n) + (u_{n-1} v_n - w_{n-1}) c(n-2,n-1) \\
&\quad + \sum_{i=1}^{n-3} v_n v_{n-1} \cdots v_{i+3} (u_{i+1} v_{i+2} - w_{i+1}) c(i,i+1) + v_n v_{n-1} \cdots v_3 \rho.
\end{aligned}$$

Lemma 2. *Let $\tau := u_0(u_1 v_2 - w_1)$. Then, for any $n \geq 4$, we have:*

$$\begin{aligned}
c(n,n-1) &= u_n c(n-1,n-1) + (u_{n-1} v_n - w_{n-1}) c(n-2,n-2) \\
&\quad + \sum_{i=1}^{n-3} v_n v_{n-1} \cdots v_{i+3} (u_{i+1} v_{i+2} - w_{i+1}) c(i,i) + v_n v_{n-1} \cdots v_3 \tau.
\end{aligned}$$

Lemma 3. *For any $n \geq 5$ and $0 \leq i \leq n - 2$, we have:*

$$\begin{aligned}
c(n,i) &= u_n c(n-1,i) + (u_{n-1} v_n - w_{n-1}) c(n-2,i-1) \\
&\quad + \sum_{j=1}^{n-3} v_n v_{n-1} \cdots v_{j+3} (u_{j+1} v_{j+2} - w_{j+1}) c(j, j+i-n+1) \\
&\quad + v_n v_{n-1} \cdots v_5 c(i-n+5,0) (u_{i-n+6} v_{i-n+7} - w_{i-n+6}).
\end{aligned}$$

To detect the positivity of $d_n(t)$, we need the following concept.

Definition 1. *Let $\alpha : \alpha_0, \alpha_1, \ldots$ be a positive weight sequence. We say that W_α is positively quadratically hyponormal if $c(n, i) \geq 0$ for all $n, i \geq 0$ with $0 \leq i \leq n + 1$, and $c(n, n + 1) > 0$ for all $n \geq 0$ [2].*

Positive quadratic hyponormality implies quadratic hyponormality, but the converse is false [15]. In addition, the authors in [15] showed that the positive quadratic hyponormality is equivalent to the quadratic hyponormality for recursively-generated weighted shift W_α with $\alpha : \sqrt{x}, \left(\sqrt{a}, \sqrt{b}, \sqrt{c} \right)^\wedge$ (here, $0 < x \leq a < b < c$).

3. Recursive Relation of $W_{\hat{\alpha}_{[5]}}$

Given the initial segment of weights $\alpha : 1, \sqrt{a}, \sqrt{b}, \sqrt{c}, \sqrt{d}$ with $1 < a < b < c < d$, we obtain the moments:

$$\gamma_0 = \gamma_1 = 1, \quad \gamma_2 = a, \quad \gamma_3 = ab, \quad \gamma_4 = abc, \quad \gamma_5 = abcd.$$

Let:

$$V_0 = \begin{pmatrix} \gamma_0 \\ \gamma_1 \\ \gamma_2 \end{pmatrix}, \quad V_1 = \begin{pmatrix} \gamma_1 \\ \gamma_2 \\ \gamma_3 \end{pmatrix}, \quad V_2 = \begin{pmatrix} \gamma_2 \\ \gamma_3 \\ \gamma_4 \end{pmatrix},$$

and we assume that V_0, V_1, V_2 are linearly independent, i.e.,

$$\det (V_0, V_1, V_2) = \det A(0, 2) \neq 0.$$

Then, there exist three nonzero numbers $\varphi_0, \varphi_1, \varphi_2$, such that:

$$\begin{pmatrix} \gamma_3 \\ \gamma_4 \\ \gamma_5 \end{pmatrix} = \varphi_0 \begin{pmatrix} \gamma_0 \\ \gamma_1 \\ \gamma_2 \end{pmatrix} + \varphi_1 \begin{pmatrix} \gamma_1 \\ \gamma_2 \\ \gamma_3 \end{pmatrix} + \varphi_2 \begin{pmatrix} \gamma_2 \\ \gamma_3 \\ \gamma_4 \end{pmatrix}.$$

A straightforward calculation shows that:

$$\begin{aligned}
\varphi_0 &= \frac{ab(ab^2 - 2abc + bc^2 + acd - bcd)}{a^2 - 2ab + ab^2 + bc - abc}, \\
\varphi_1 &= \frac{-ab(ab - ac - bc + bc^2 + cd - bcd)}{a^2 - 2ab + ab^2 + bc - abc}, \\
\varphi_2 &= \frac{b(a^2 - ab - ac + abc + cd - acd)}{a^2 - 2ab + ab^2 + bc - abc}.
\end{aligned}$$

Thus:

$$\gamma_{n+1} = \varphi_0 \gamma_{n-2} + \varphi_1 \gamma_{n-1} + \varphi_2 \gamma_n \qquad (n \geq 5),$$

i.e.,

$$\alpha_n^2 = \varphi_2 + \frac{\varphi_1}{\alpha_{n-1}^2} + \frac{\varphi_0}{\alpha_{n-1}^2 \alpha_{n-2}^2} \qquad (n \geq 5). \tag{1}$$

By (1), we can obtain a recursively-generated weighted shift, and we set it as $\hat{\alpha}_{[5]}$. In this case, we call the weighted shift operator $W_{\hat{\alpha}_{[5]}}$ with rank three.

Proposition 1. *$W_{\hat{\alpha}_{[5]}}$ with rank three is subnormal if and only if:*
(1) $1 < a < b$,
(2) $c > \frac{a}{b} \left(\frac{(b-1)^2}{a-1} + 1 \right)$,
(3) $d > \frac{b}{c} \left(\frac{(c-a)^2}{b-a} + a \right)$.

Proof. See [16], Example 3.6. $\square$

Proposition 2. *If $W_{\hat{\alpha}_{[5]}}$ with rank three is subnormal, then $\varphi_0 > 0$, $\varphi_1 < 0$ and $\varphi_2 > 0$.*

Proof. By Proposition 1, we know that:

$$
\begin{aligned}
F_1 \ &: \ = ab^2 - 2ab + bc + a^2 - abc < 0, \\
F_2 \ &: \ = ab^2 + bc^2 - 2abc + acd - bcd < 0.
\end{aligned}
$$

Thus, $\varphi_0 > 0$. Since:

$$
ab - ac - bc + bc^2 + c\,(1 - b)\,d < (c - b)\,\frac{F_1}{b - a} < 0,
$$

and:

$$
a^2 - ab - ac + abc + c\,(1 - a)\,d < (c - a)\,\frac{F_1}{b - a} < 0,
$$

we have $\varphi_1 < 0$ and $\varphi_2 > 0$. The proof is complete. $\square$

Proposition 3. *Let $\frac{u_{n-1}}{u_n} = \beta_n \ (n \geq 5)$. Then:*

$$
\beta_n = -\frac{\alpha_{n-1}^2 \alpha_{n-2}^2 \alpha_{n-3}^2}{\varphi_1 \alpha_{n-3}^2 + \varphi_0 + \varphi_0 \beta_{n-1}}. \tag{2}
$$

Proof. Since:

$$
\begin{aligned}
u_n \ &= \ \alpha_n^2 - \alpha_{n-1}^2 \\
&= \ -\frac{\varphi_1}{\alpha_{n-1}^2 \alpha_{n-2}^2} u_{n-1} - \frac{\varphi_0}{\alpha_{n-1}^2 \alpha_{n-2}^2 \alpha_{n-3}^2}\,(u_{n-1} + u_{n-2}) \\
&= \ -\left(\frac{\varphi_1}{\alpha_{n-1}^2 \alpha_{n-2}^2} + \frac{\varphi_0}{\alpha_{n-1}^2 \alpha_{n-2}^2 \alpha_{n-3}^2} \right) u_{n-1} - \frac{\varphi_0}{\alpha_{n-1}^2 \alpha_{n-2}^2 \alpha_{n-3}^2} u_{n-2},
\end{aligned}
$$

so:

$$
\begin{aligned}
1 \ &= \ -\left(\frac{\varphi_1}{\alpha_{n-1}^2 \alpha_{n-2}^2} + \frac{\varphi_0}{\alpha_{n-1}^2 \alpha_{n-2}^2 \alpha_{n-3}^2} \right) \frac{u_{n-1}}{u_n} - \frac{\varphi_0}{\alpha_{n-1}^2 \alpha_{n-2}^2 \alpha_{n-3}^2} \frac{u_{n-2}}{u_n} \\
&= \ -\left(\frac{\varphi_1}{\alpha_{n-1}^2 \alpha_{n-2}^2} + \frac{\varphi_0}{\alpha_{n-1}^2 \alpha_{n-2}^2 \alpha_{n-3}^2} + \frac{\varphi_0}{\alpha_{n-1}^2 \alpha_{n-2}^2 \alpha_{n-3}^2} \frac{u_{n-2}}{u_{n-1}} \right) \frac{u_{n-1}}{u_n}.
\end{aligned}
$$

Thus, we have:

$$
\begin{aligned}
\beta_n \ &= \ -\frac{1}{\left(\dfrac{\varphi_1}{\alpha_{n-1}^2 \alpha_{n-2}^2} + \dfrac{\varphi_0}{\alpha_{n-1}^2 \alpha_{n-2}^2 \alpha_{n-3}^2} + \dfrac{\varphi_0}{\alpha_{n-1}^2 \alpha_{n-2}^2 \alpha_{n-3}^2} \beta_{n-1} \right)} \\
&= \ -\frac{\alpha_{n-1}^2 \alpha_{n-2}^2 \alpha_{n-3}^2}{\varphi_1 \alpha_{n-3}^2 + \varphi_0 + \varphi_0 \beta_{n-1}}.
\end{aligned}
$$

Thus, we have our conclusion. $\square$

Since $\lim_{n \to \infty} \alpha_n^2 = L^2$, we let $\lim_{n \to \infty} \beta_n := \beta^*$, and by (2), we have $\beta^* = -\dfrac{L^6}{\varphi_1 L^2 + \varphi_0 + \varphi_0 \beta^*}$; hence:

$$
\beta^* = -\frac{1}{2} - \frac{\varphi_1}{2\varphi_0} L^2 + \frac{1}{2\varphi_0} \sqrt{\left(\varphi_1^2 - 4\varphi_0 \varphi_2 \right) L^4 - 2\varphi_0 \varphi_1 L^2 - 3\varphi_0^2}. \tag{3}
$$

4. Main Results

First, we give the following result (cf. [11], Corollary 5).

Proposition 4. *Let W_α be any unilateral weighted shift. Then, W_α is two-hyponormal if and only if $\theta_k :=$ $u_k v_{k+1} - w_k \geq 0, \forall k \in \mathbb{N}$.*

It is well-known that if W_α is two-hyponormal or positively quadratically hyponormal, then W_α is quadratically hyponormal. By Proposition 4 and Lemma 1~3, we have the following result.

Theorem 1. *Let W_α be any unilateral weighted shift. If W_α is 2-hyponormal, then W_α is positively quadratically hyponormal.*

4.1. The Positive Quadratic Hyponormality of $W_{\alpha(x)}$

Let $\alpha(x) : \sqrt{x}, \hat{\alpha}_{[5]}$ with $0 < x \leq 1$, and we consider the following $(n+1) \times (n+1)$ matrix:

$$D_n = \begin{bmatrix} u_0 + v_0 t & \sqrt{w_0 t} & 0 & \cdots & 0 \\ \sqrt{w_0 t} & u_1 + v_1 t & \sqrt{w_1 t} & \cdots & 0 \\ 0 & \sqrt{w_1 t} & u_2 + v_2 t & \ddots & 0 \\ \vdots & \vdots & \ddots & \ddots & \sqrt{w_{n-1} t} \\ 0 & 0 & \cdots & \sqrt{w_{n-1} t} & u_n + v_n t \end{bmatrix}. \tag{4}$$

Let $d_n = \det D_n = \sum_{i=0}^{n+1} c(n,i) t^i$. Then:

Lemma 4. $c(n,i) \geq 0, n = 0, 1, 2,$ and $0 \leq i \leq n+1$.

Proof. In fact, $c(1,1) = x(a-x) > 0$, $c(2,1) = ax(1-x)(b-1) > 0$, and:

$$\begin{aligned} c(2,2) &= ax((ab-1) - (b-1)x) \\ &> ax((ab-a) - (b-1)x) \\ &= ax(a-x)(b-1) > 0. \end{aligned}$$

Thus, we have our conclusion. $\square$

Lemma 5. *Assume that $\theta_k := u_k v_{k+1} - w_k \geq 0$ for $k \geq 2$. Then, for $n \geq 3, 0 \leq i \leq n+1$, we have:*

$$c(n,i) \geq u_n c(n-1,i) + v_n \cdots v_3[v_2 c(1,i-n+1) - w_1 c(0,i-n+1)].$$

Proof. For $n = 3, 0 \leq i \leq 4$,

$$\begin{aligned} c(3,i) &= u_3 c(2,i) + v_3 c(2,i-1) - w_2 c(1,i-1) \\ &= u_3 c(2,i) + (\mathbf{u_2 v_3 - w_2}) c(1,i-1) + v_3[v_2 c(1,i-2) - w_1 c(0,i-2)] \\ &\geq u_3 c(2,i) + v_3[v_2 c(1,i-2) - w_1 c(0,i-2)]. \end{aligned}$$

By the inductive hypothesis, we have our result. $\square$

Thus, if $\theta_k := u_k v_{k+1} - w_k \geq 0$ for $k \geq 2$, then by Lemma 1~3 and Lemma 5, for $n \geq 3$, we have:

$$c(n,i) \geq \begin{cases} v_n \cdots v_2 c(1,2), & \text{for} \quad i = n+1, \\ u_n c(n-1,n) + v_n \cdots v_3 \rho, & \text{for} \quad i = n, \\ u_n c(n-1,n-1) + v_n \cdots v_3 \tau, & \text{for} \quad i = n-1, \\ u_n c(n-1,i), & \text{for} \quad 0 \leq i \leq n-2, \end{cases}$$

where $\rho = ax(b-1)(a-x) > 0$, $\tau = x(a(b-a) - (ab-2a+1)x)$. Therefore, when $n \geq 3$, we have:

$$\begin{cases} c(n,n+1) > 0, \\ c(n,n) > u_n c(n-1,n) + v_n \cdots v_3 \rho \geq 0, \\ c(n,i) \geq u_n \cdots u_{i+2} c(i+1,i) \quad (n \geq 3, 0 \leq i \leq n-2). \end{cases}$$

To complete our analysis of the coefficients $c(n,i)$, it suffices to determine the values of x for which $c(n,n-1) \geq 0$ $(n \geq 3)$.

Lemma 6. $c(n,n-1) \geq 0$ *(for any $n \geq 3$), if $c(3,2) \geq 0, c(4,3) \geq 0$ and $A_n := v_0 v_1 v_2 u_n u_{n-1} + u_n v_{n-1} \rho + v_n v_{n-1} \tau \geq 0$ $(n \geq 5)$.*

Proof. For $n \geq 4$, by Lemma 2, we have:

$$\begin{aligned} c(n,n-1) &\geq u_n c(n-1,n-1) + v_n \cdots v_3 \tau \\ &\geq u_n [u_{n-1} c(n-2,n-1) + v_{n-1} \cdots v_3 \rho] + v_n \cdots v_3 \tau, \end{aligned}$$

and since $c(n-2,n-1) = v_{n-2} \cdots v_0$, we get:

$$c(n,n-1) \geq u_n(u_{n-1} v_{n-2} \cdots v_0 + v_{n-1} \cdots v_3 \rho) + v_n \cdots v_3 \tau.$$

If $n \geq 5$, we can factor $v_{n-2} \cdots v_3$ to get:

$$\begin{aligned} c(n,n-1) &\geq v_{n-2} \cdots v_3(v_0 v_1 v_2 u_n u_{n-1} + u_n v_{n-1} \rho + v_n v_{n-1} \tau) \\ &= v_{n-2} \cdots v_3 A_n. \end{aligned}$$

Hence, we have our result. $\square$

Let:

$$x_n := \sup\{x : c(n,n-1) \geq 0 \text{ in } W_\alpha\} \quad (n \geq 3).$$

By direct computations, we have:

$$\begin{aligned} x_3 &= \frac{a\theta_2 + a(b-a)v_3 + a(ab-1)u_3}{\theta_2 + a(b-1)u_3 + (-2a+ab+1)v_3}, \\ x_4 &= \frac{a(ab-1)\theta_3 + a(u_4+v_4)\theta_2 + (a^2(b-1)u_4 + v_4 a(b-a))v_3 + a^2 b u_3 u_4}{a(b-1)\theta_3 + v_4 \theta_2 + ((ab-2a+1)v_4 + u_4 a(b-1))v_3 + a u_3 u_4}. \end{aligned}$$

For $n \geq 5$, a calculation using the specific form of v_0, v_1, v_2, ρ and τ shows that:

$$\begin{aligned} A_n &= [a^2 b u_n u_{n-1} + a^2(b-1)u_n v_{n-1} + a(b-a)v_n v_{n-1} \\ &\quad - (a u_n u_{n-1} + a(b-1)u_n v_{n-1} + (ab+1-2a)v_n v_{n-1})x]x, \end{aligned}$$

it follows that:

$$x_n = \frac{a^2 b u_n u_{n-1} + a^2(b-1)u_n v_{n-1} + a(b-a)v_n v_{n-1}}{a u_n u_{n-1} + a(b-1)u_n v_{n-1} + (ab+1-2a)v_n v_{n-1}}.$$

Let $z_n := \frac{v_n}{u_n}$ $(u_n \neq 0)$. Then, for $n \geq 5$,

$$x_n = \frac{a^2 b + a^2(b-1)z_{n-1} + a(b-a)z_n z_{n-1}}{a + a(b-1)z_{n-1} + (ab+1-2a)z_n z_{n-1}}. \tag{5}$$

Lemma 7. $\lim_{n\to\infty} z_n = K := (1+\beta^*)\left(\varphi_2 - \frac{\varphi_0}{L^4}\beta^*\right)$, *where β^* as in (3).*

Proof. Since:

$$\alpha_{n+1}^2 = \varphi_2 + \frac{\varphi_1}{\alpha_n^2} + \frac{\varphi_0}{\alpha_n^2 \alpha_{n-1}^2} \quad (n \geq 5),$$

from which it follows that:

$$\alpha_n^2 \alpha_{n+1}^2 = \varphi_2 \alpha_n^2 + \varphi_1 + \frac{\varphi_0}{\alpha_{n-1}^2}.$$

Thus:

$$v_n = \varphi_2(u_n + u_{n-1}) - \frac{\varphi_0(u_{n-1} + u_{n-2})}{\alpha_{n-1}^2 \alpha_{n-3}^2}. \quad (n \geq 5) \tag{6}$$

Thus, we have (for $n \geq 5$):

$$z_n = \varphi_2(1+\beta_n) - \frac{\varphi_0}{\alpha_{n-1}^2 \alpha_{n-3}^2}\beta_n(1+\beta_{n-1}).$$

Since $\alpha_n^2 \to L^2$, we have:

$$\lim_{n\to\infty} z_n = (1+\beta^*)\left(\varphi_2 - \frac{\varphi_0}{L^4}\beta^*\right).$$

Thus, we have our conclusion. $\square$

Let:

$$f(z,w) := \frac{a^2 b + a^2(b-1)z + a(b-a)zw}{a + a(b-1)z + (ab+1-2a)zw}.$$

By Lemma 7 and the fact in [2] (p. 399), if z_n is increasing, then we know that $\{x_n\}_{n\geq 5}$ in (5) is decreasing and $\inf_{n\geq 5} x_n = f(K,K)$. Thus, we have the following result.

Theorem 2. *Assume that $W_{\hat{\alpha}_{[5]}}$ with rank three is subnormal. Let $\alpha\,(x) : \sqrt{x}, \hat{\alpha}_{[5]}$, and let:*

$$h_2^+ := \sup\{x : W_{\alpha(x)} \text{ be positively quadratically hyponormal}\}.$$

If $z_n := \frac{v_n}{u_n}$ $(n \geq 5)$ is increasing, then:

$$h_2^+ \geq \min\left\{1, x_3, x_4, \frac{a^2 b + a^2(b-1)K + a(b-a)K^2}{a + a(b-1)K + (ab+1-2a)K^2}\right\},$$

where:

$$x_3 = \frac{a\theta_2 + a\,(b-a)\,v_3 + a\,(ab-1)\,u_3}{\theta_2 + a\,(b-1)\,u_3 + (-2a+ab+1)\,v_3},$$

$$x_4 = \frac{a\,(ab-1)\,\theta_3 + a\,(u_4+v_4)\,\theta_2 + \left(a^2\,(b-1)\,u_4 + a\,(b-a)\,v_4\right)v_3 + a^2 b u_3 u_4}{a\,(b-1)\,\theta_3 + v_4\theta_2 + (a\,(b-1)\,u_4 + (ab-2a+1)\,v_4)\,v_3 + a u_3 u_4},$$

$$K = (1+\beta^*)\left(\varphi_2 - \frac{\varphi_0}{L^4}\beta^*\right).$$

Remark 1. *By (6), we know that if β_n is increasing, then so is z_n. Hence, our problems are as follows.*

Problem 1. *Let $\alpha := \{\alpha_n\}_{n=0}^{\infty}$ be any unilateral weighted sequence. If W_α is subnormal, is β_n increasing or not? In particular, what is the answer for subnormal $W_{\hat{\alpha}_{[5]}}$ with rank three?*

Example 1. *Let $\alpha(x) : \sqrt{x}, \left(1, \sqrt{2}, \sqrt{3}, \sqrt{4}, \sqrt{5}\right)^{\wedge}$. Then, $\varphi_0 = 6, \varphi_1 = -18, \varphi_2 = 9, L^2 \approx 6.2899$, and $K \approx 32.118, x_3 = \frac{17}{18} \approx 0.94444, x_4 = \frac{226}{249} \approx 0.90763, f(K, K) \approx 0.77512$. We obtain $h_2^{+} \gtrsim 0.77512$. That is, if $0 < x \lesssim 0.77512$, then $W_{\alpha(x)}$ is positively quadratically hyponormal. Numerically, we know that β_n and z_n are all increasing. See the following Table 1.*

Table 1. Numerical data for β_n and z_n in Example 1.

n	7	8	9	10	11	12
β_n	1.9851	2.3929	2.6008	2.6882	2.7218	2.7343
z_n	25.597	29.120	30.909	31.660	31.949	32.056
n	13	14	15	16	17	18
β_n	2.7389	2.7406	2.7412	2.7414	2.7415	2.7415
z_n	32.096	32.11	32.116	32.117	32.118	32.118

4.2. The Positive Quadratic Hyponormality of $W_{\alpha(y,x)}$

Let $\alpha(y, x) : \sqrt{y}, \sqrt{x}, \hat{\alpha}_{[5]}$. We also consider the matrix as in (4), and let $d_n = \det D_n = \sum_{i=0}^{n+1} c(n, i)\, t^i$. Then:

$$
\begin{aligned}
c(1,1) &= xy(1-y) > 0, \\
c(2,1) &= y(x-y)(a-x) > 0, \\
c(2,2) &= xy\left(a - ay + xy - x^2\right) \geq xy(1-y)(a-x) > 0,
\end{aligned}
$$

and:

$$
\begin{aligned}
c(3,1) &\geq u_3 c(2,1) > 0, \\
c(3,2) &= y\left((x-a)\left(x^2 - 2x + ab\right) y + x\left((1-a)x^2 + a(1-b)x + a(ab-1)\right)\right), \\
c(3,3) &\geq u_3 c(2,3) + v_3\rho > 0.
\end{aligned}
$$

Since:

$$
\begin{aligned}
\rho &= xy(1-y)(a-x) > 0, \\
\tau &= y\left(\left(2x - x^2 - a\right)y + x(a-1)\right),
\end{aligned}
$$

we can similarly show that $c(n, n-1) \geq 0$ for all $n \geq 3$, if $c(3,2) \geq 0, c(4,3) \geq 0$ and $y \leq \frac{x\left((a-1)K^2 + (a-x)K + ax\right)}{(a-2x+x^2)K^2 + (a-x)xK + x^3}$, where $K = (1 + \beta^*)\left(\varphi_2 - \frac{\varphi_0}{L^4}\beta^*\right)$. Thus, we have the following result.

Theorem 3. *Let $\alpha(y, x) : \sqrt{y}, \sqrt{x}, \hat{\alpha}_{[5]}$. If $z_n := \frac{v_n}{u_n}$ $(n \geq 5)$ is increasing, and:*

$$(1)\ 0 < x \leq \min\left\{a - \sqrt{a(a-1)},\ \frac{ab(c-1) - \sqrt{ab(a-1)(a^2b + bc^2 - ab - ac + 2a^2 - a^3 - abc)}}{a(a-1) + b(c-a)}\right\},$$

(2) $0 < y \leq \min\{x, f_1(x), f_2(x), f_3(x)\}$, *where*:

$$f_1(x) = \frac{x\left(a(ab-1)+a(1-b)x-(a-1)x^2\right)}{(a-x)(ab-2x+x^2)},$$

$$f_2(x) = -\frac{xp_1(x)}{p_2(x)}, \text{ with:}$$

$$p_1(x) = (a^2b+bc-a^2-abc)x^2 + (a^2b-2ab+a^2-ab^2c+abc)x$$
$$+ (ca^2b^2 - 2a^3b + 2a^2b - cab),$$
$$p_2(x) = (bc-ab-a+a^2)x^3 + (a-a^2b+3ab-2bc-abc)x^2$$
$$+ (a^3b - 3a^2b - a^2 + cab^2 + 2cab)x + a^2b(a-bc),$$

$$f_3(x) = \frac{x\left((a-1)K^2+(a-x)K+ax\right)}{(a-2x+x^2)K^2+(a-x)xK+x^3}, \quad K = (1+\beta^*)\left(\varphi_2 - \frac{\varphi_0}{L^4}\beta^*\right),$$

then $W_{\alpha(y,x)}$ *is positively quadratically hyponormal.*

Example 2. *Let* $a = 2, b = 3, c = 4, d = 5$. *If* $0 < x \leq 2 - \sqrt{2} \approx 0.58579$, $y \leq \frac{x\left(K^2+(2-x)K+2x\right)}{(x^2-2x+2)K^2+(2x-x^2)K+x^3}$ *with* $K \approx 32.118$, *then* $W_{\alpha(y,x)}$ *is positively quadratically hyponormal. See the following Figure* 1.

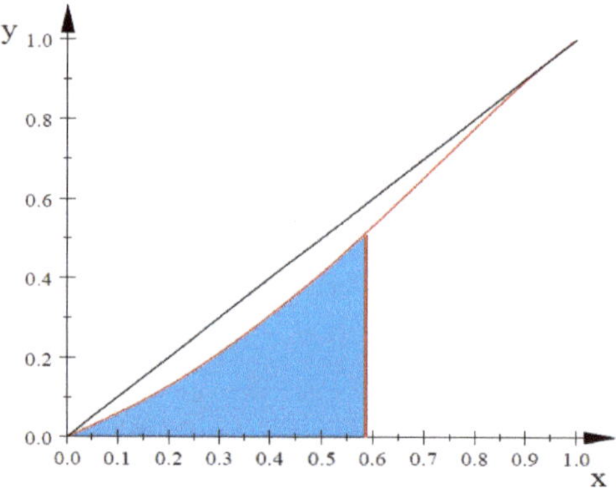

Figure 1. A subset of the region of positive quadratic hyponormality of $W_{\alpha(y,x)}$ in Example 2.

5. More Results

From the above discussions, we obtain the following criteria for any unilateral weighted shifts.

Proposition 5. *Let* $\alpha(x) : \sqrt{x}, 1, \sqrt{a}, \sqrt{b}, \sqrt{c}, \sqrt{d}, \sqrt{\alpha_5}, \sqrt{\alpha_6}, \ldots,$ *and* $\alpha : 1, \sqrt{a}, \sqrt{b}, \sqrt{c}, \sqrt{d}, \sqrt{\alpha_5}, \sqrt{\alpha_6}, \ldots$ *be subnormal weighted shifts. Let:*

$$h_2^+ := \sup\{x : W_{\alpha(x)} \text{ be positively quadratically hyponormal}\}.$$

If $z_n = \frac{v_n}{u_n}$ $(n \geq 5)$ *is increasing and* $z_n \to K$ *(as* $n \to \infty$*), then:*

$$h_2^+ \geq \min\left\{1, x_3, x_4, \frac{a^2b+a^2(b-1)K+a(b-a)K^2}{a+a(b-1)K+(ab+1-2a)K^2}\right\},$$

where:

$$x_3 = \frac{a\theta_2 + a\,(b-a)\,v_3 + a\,(ab-1)\,u_3}{\theta_2 + a\,(b-1)\,u_3 + (-2a+ab+1)\,v_3},$$

$$x_4 = \frac{a\,(ab-1)\,\theta_3 + a\,(u_4+v_4)\,\theta_2 + \left(a^2\,(b-1)\,u_4 + a\,(b-a)\,v_4\right)v_3 + a^2 b u_3 u_4}{a\,(b-1)\,\theta_3 + v_4\theta_2 + \left(a\,(b-1)\,u_4 + (ab-2a+1)\,v_4\right)v_3 + a u_3 u_4}.$$

By Proposition 5, we can have the following results, but we omit the concrete computations.

Example 3. *(1) Let* $\alpha(x) : \sqrt{x}, \sqrt{\frac{2}{3}}, \sqrt{\frac{3}{4}}, \sqrt{\frac{4}{5}}, \ldots$ *Then,* $h_2^+ = \frac{2}{3}$ *(cf. [11], Proposition 7).*

(2) Let $\alpha(x) : \sqrt{x}, \sqrt{\frac{5}{8}}, \sqrt{\frac{3}{4}}, \sqrt{\frac{4}{5}}, \ldots$ *Then,* $h_2^+ = \frac{1945}{3136}$ *(cf. [15], Theorem 3.7).*

(3) Let $\alpha(x) : \sqrt{x}, 1, \left(\sqrt{2}, \sqrt{2.1}, \sqrt{12.1}\right)^{\wedge}$. *Then,* $h_2^+ \gtrsim 0.16682$.

(4) Let $\alpha(x) : \sqrt{x}, \sqrt{\frac{n}{n+1} \cdot \frac{1}{2} \cdot \frac{2^{n+1}-1}{2^n-1}}$. *Then,* $h_2^+ = \frac{3}{4}$ *(cf. [16], Example 3.4).*

6. Conclusions

In this work, we study a weighed shift operator for which the weights are recursively generated by five weights. We give sufficient conditions of the positive quadratic hyponormalities. Next, it is worth studying the cubic hyponormality, semi-weak k-hyponormalities, and so on.

Author Contributions: All authors contributed equally in writing this article. All authors read and approved the final manuscript.

Funding: This work was supported by the National Research Foundation of Korea (NRF) grant funded by the Korea government (MEST) (No. 2017R1A2B4006092).

Conflicts of Interest: The authors declare no conflict of interest.

References

1. Baek, S.; Exner, G.; Jung, I.B.; Li, C. Semi-cubic hyponormality of weighted shifts with Stampfli recursive tail. *Integral Equ. Oper. Theory* **2017**, *88*, 229–248. [CrossRef]
2. Curto, R.; Fialkow, L. Recursively generated weighted shifts and the subnormal completion problem, II. *Integral Equ. Oper. Theory* **1994**, *18*, 369–426. [CrossRef]
3. Curto, R.; Jung, I.B. Quadratically hyponormal weighted shifts with first two equal weights. *Integral Equ. Oper. Theory* **2000**, *37*, 208–231. [CrossRef]
4. Dong, Y.; Exner, G.; Jung, I.B.; Li, C. Quadratically hyponormal recursively generated weighted shifts. *Oper. Theory Adv. Appl.* **2008**, *187*, 141–155.
5. Exner, G.; Jung, I.B.; Park, S.S. Weakly n-hyponormal weighted shifts and their examples. *Integral Equ. Oper. Theory* **2006**, *54*, 215–233. [CrossRef]
6. Li, C.; Lee, M.R. Existence of non-subnormal completely semi-weakly hyponormal weighted shifts. *Filomat* **2017**, *31*, 1627–1638. [CrossRef]
7. Jung, I.B.; Park, S.S. Cubically hyponormal weighted shifts and their examples. *J. Math. Anal. Appl.* **2000**, *247*, 557–569. [CrossRef]
8. Li, C.; Cho, M.; Lee, M.R. A note on cubically hyponormal weighted shifts. *Bull. Korean Math. Soc.* **2014**, *51*, 1031–1040. [CrossRef]
9. Exner, G.; Jin, J.Y.; Jung, I.B.; Lee, J.E. Weak Hamburger-type weighted shifts and their examples. *J. Math. Anal. Appl.* **2018**, *462*, 1357–1380. [CrossRef]
10. Curto, R.; Fialkow, L. Recursively generated weighted shifts and the subnormal completion problem. *Integral Equ. Oper. Theory* **1993**, *17*, 202–246. [CrossRef]
11. Curto, R. Quadratically hyponormal weighted shifts. *Integral Equ. Oper. Theory* **1990**, *13*, 49–66. [CrossRef]
12. Curto, R.; Putinar, M. Existence of non-subnormal polynomially hyponormal operators. *Bull. Am. Math. Soc.* **1991**, *25*, 373–378. [CrossRef]
13. Curto, R.; Putinar, M. Nearly subnormal operators and moment problems. *J. Funct. Anal.* **1993**, *115*, 480–497. [CrossRef]

14. Li, C.; Lee, M.R.; Baek, S.H. A relationship: Subnormal, polynomially hyponormal and semi-weakly hyponormal weighted shifts. *J. Math. Anal. Appl.* **2019**, Under review.

15. Jung, I.B.; Park, S.S. Quadratically hyponormal weighted shifts and their examples. *Integral Equ. Oper. Theory* **2002**, *36*, 480–498. [CrossRef]

16. Jung, I.B.; Li, C. A formula for k-hyponormality of backstep extensions of subnormal weighted shifts. *Proc. Am. Math. Soc.* **2001**, *129*, 2343–2351. [CrossRef]

 *mathematics*

Article

Ground State Solutions for Fractional Choquard Equations with Potential Vanishing at Infinity

Huxiao Luo [1], Shengjun Li [2,*] and Chunji Li [3]

[1]   Department of Mathematics, Zhejiang Normal University, Jinhua 321004, China; luohuxiao@zjnu.edu.cn
[2]   College of Information Sciences and Technology, Hainan University, Haikou 570228, China
[3]   Department of Mathematics, Northeastern University, Shenyang 110004, China; lichunji@mail.neu.edu.cn
*   Correspondence: shjli626@126.com

Received: 8 December 2018; Accepted: 1 February 2019; Published: 5 February 2019

Abstract: In this paper, we study a class of nonlinear Choquard equation driven by the fractional Laplacian. When the potential function vanishes at infinity, we obtain the existence of a ground state solution for the fractional Choquard equation by using a non-Nehari manifold method. Moreover, in the zero mass case, we obtain a nontrivial solution by using a perturbation method. The results improve upon those in Alves, Figueiredo, and Yang (2015) and Shen, Gao, and Yang (2016).

Keywords: variational methods; fractional Choquard equation; ground state solution; vanishing potential

MSC: 35J50; 58E30

---

1. Introduction

In this paper, we deal with the following nonlocal equation:

$$\begin{cases} (-\Delta)^s u + V(x)u = \left(\int_{\mathbb{R}^N} \frac{Q(y)F(u(y))}{|x-y|^\mu} dy \right) Q(x)f(u), & \text{in } \mathbb{R}^N, \\ u \in D^{s,2}(\mathbb{R}^N), \end{cases} \tag{1}$$

where $N \geq 3$, $0 < s < 1$, $0 < \mu < N$, $V \in C(\mathbb{R}^N, [0, \infty))$, $Q \in C(\mathbb{R}^N, (0, \infty))$, $f \in C(\mathbb{R}, \mathbb{R})$ and $F(t) = \int_0^t f(s)ds$. The fractional Laplacian $(-\Delta)^s$ is defined as

$$(-\Delta)^s u(x) = C_{N,s} P.V. \int_{\mathbb{R}^N} \frac{u(x) - u(y)}{|x-y|^{N+2s}} dy, \quad u \in \mathcal{S}(\mathbb{R}^N),$$

where P.V. denotes the *principal value* of the singular integral, $\mathcal{S}(\mathbb{R}^N)$ is the Schwartz space of rapidly decaying C^∞ functions in $\mathbb{R}^N$, and

$$C_{N,s} = \frac{2^{2s} s \Gamma(N+s)}{\pi^{N/2} \Gamma(1-s)}.$$

$(-\Delta)^s$ is a pseudo-differential operator, and can be equivalently defined via Fourier transform as

$$\mathscr{F}[(-\Delta)^s u](\xi) = |\xi|^{2s} \mathscr{F}[u](\xi), \quad u \in \mathcal{S}(\mathbb{R}^N),$$

Mathematics **2019**, *7*, 151; doi:10.3390/math7020151   www.mdpi.com/journal/mathematics

where $\mathscr{F}$ is the Fourier transform, that is,

$$\mathscr{F}[u](\xi) = \frac{1}{(2\pi)^{\frac{N}{2}}} \int_{\mathbb{R}^N} e^{-i\xi \cdot x} u(x) dx, \ u \in \mathcal{S}(\mathbb{R}^N).$$

The fractional Laplace operator $(-\Delta)^s$ is the infinitesimal generator of Lévy stable diffusion processes, and appears in several areas such as the thin obstacle problem, anomalous diffusion, optimization, finance, phase transitions, crystal dislocation, multiple scattering, and materials science, see [1–5] and their references.

Recently, a great deal of work has been devoted to the study of the Choquard equations, see [6–14] and their references. For instance, Alves, Cassani, Tarsi, and Yang [7] studied the following singularly perturbed nonlocal Schrödinger equation:

$$-\varepsilon^2 \Delta u + V(x)u = \varepsilon^{\mu-2} \left[\frac{1}{|x|^\mu} * F(u) \right] f(u), \ \text{in } \mathbb{R}^2,$$

where $0 < \mu < 2$ and ε is a positive parameter, the nonlinearity f has critical exponential growth in the sense of Trudinger–Moser. By using variational methods, the authors established the existence and concentration of solutions for the above equation.

In [6], Alves, Figueiredo and Yang studied the following Choquard equation:

$$\begin{cases} -\Delta u + V(x)u = (\frac{1}{|x|^\mu} * F(u))f(u), \ \text{in } \mathbb{R}^N. \\ u \in H^1(\mathbb{R}^N). \end{cases} \tag{2}$$

Under the assumption $V(x) \to 0$ as $|x| \to \infty$, the authors obtained a nontrivial solution for (2) by using a penalization method.

In the physical case $N = 3, \mu = 1, V(x) = 1$ and $F(t) = \frac{t^2}{2}$, (2) is also known as the stationary Hartree equation [15]. It dates back to the description of the quantum mechanics of a polaron at rest by Pekar in 1954 [16]. In 1976, Choquard used (2) to describe an electron trapped in its own hole, in a certain approximation to the Hartree–Fock theory of one-component plasma [11]. In 1996, Penrose proposed (2) as a model of self-gravitating matter, in a programme in which quantum state reduction is understood as a gravitational phenomenon [15].

In addition, there is little literature on the fractional Choquard equations. Frank and Lenzmann [17] established the uniqueness and radial symmetry of ground state solutions for the following equation:

$$(-\Delta)^{\frac{1}{2}} u + u = (|x|^{-1} * |u|^2)u, \ \text{in } \mathbb{R}^N.$$

D'Avenia, Siciliano, and Squassina [18] obtained the existence, regularity, symmetry, and asymptotic of the solutions for the nonlocal problem

$$(-\Delta)^s u + \omega u = (|x|^{-\mu} * |u|^p)|u|^{p-2}u, \ \text{in } \mathbb{R}^N.$$

In [19], Shen, Gao, and Yang studied the following fractional Choquard equation:

$$(-\Delta)^s u + u = (|x|^{-\mu} * F(u))f(u), \ \text{in } \mathbb{R}^N, \tag{3}$$

where $N \geq 3, s \in (0,1)$, and $\mu \in (0,N)$. Under the general Berestycki–Lions-type conditions [20], the authors obtained the existence and regularity of ground states for (3). The authors also established the Pohožaev identity for (3):

$$\frac{N-2s}{2} \int_{\mathbb{R}^N} |(-\Delta)^{\frac{s}{2}} u|^2 dx + \frac{N}{2} \int_{\mathbb{R}^N} u^2 dx = \frac{2N-\mu}{2} \int_{\mathbb{R}^N} (|x|^{-\mu} * F(u)) F(u) dx.$$

Motivated by the above works, in the first part of this article, we study the ground state solution for (1). We assume

(I) $V(x), Q(x) > 0$ for all $x \in \mathbb{R}^N, V \in C(\mathbb{R}^N, \mathbb{R})$ and $Q \in C(\mathbb{R}^N, \mathbb{R}) \cap L^\infty(\mathbb{R}^N, \mathbb{R})$;

(II)   if $\{A_n\} \subset \mathbb{R}^N$ is a sequence of Borel sets such that $\mathrm{meas}\{A_n\} \leq \delta$ for all n and some $\delta > 0$, then

$$\lim_{r \to \infty} \int_{A_n \cap B_r^c(0)} [Q(x)]^{\frac{2N}{2N-\mu}} dx = 0 \text{ uniformly in } n \in \mathbb{N};$$

(III)   one of the below conditions occurs:

$$\frac{Q}{V} \in L^\infty(\mathbb{R}^N), \tag{4}$$

or there exists $p \in (2, 2_s^*)$ such that

$$\frac{[Q(x)]^{\frac{2N}{2N-\mu}}}{[V(x)]^{\frac{2_s^*-p}{2_s^*-2}}} \to 0, \quad |x| \to \infty, \tag{5}$$

where $2_s^* = \frac{2N}{N-2s}$ is the fractional critical exponent;

(F1) $F(t) = o(|t|^{\frac{2N-\mu}{N}})$ as $t \to 0$ if (4) holds; or $F(t) = o(|t|^{\frac{p(2N-\mu)}{2N}})$ as $t \to 0$ if (5) holds;

(F2) $F(t) = o(|t|^{\frac{2N-\mu}{N-2s}})$ as $t \to \infty$;

(F3) $f(t)$ is nondecreasing on $\mathbb{R}$;

(F4) $\lim\limits_{|t| \to +\infty} \frac{F(t)}{|t|} = +\infty$.

It is necessary for us to point out that the original of assumptions (I)–(III) come from [21–23]. The assumptions can be used to prove that the work space E is compactly embedded into the weighted Lebesgue space $L_K^q(\mathbb{R}^N)$, see Section 2 and Lemma 1.

Now, we can state the first result of this article.

Theorem 1. *Suppose that* $(I), (II), (III)$ *and (F1)–(F4) hold. Then* (1) *has a ground state solution.*

Remark 1. *Since the Nehari-type monotonicity condition for f is not satisfied, the Nehari manifold method used in [24] no longer works in our setting. To prove Theorem 2, we use the non-Nehari manifold method developed by Tang [25], which relies on finding a minimizing sequence outside the Nehari manifold by using the diagonal method (see Lemma 8).*

In the second part of this article, we consider the following fractional Choquard equation with zero mass case:

$$\begin{cases} (-\Delta)^s u = \left(\frac{1}{|x|^\mu} * F(u) \right) f(u), & \text{in } \mathbb{R}^N, \\ u \in D^{s,2}(\mathbb{R}^N), \end{cases} \tag{6}$$

where $N \geq 3, 0 < s < 1, 0 < \mu < \min\{N, 4s\}$. The homogeneous fractional Sobolev space $D^{s,2}(\mathbb{R}^N)$, also denoted by $\dot{H}^s(\mathbb{R}^N)$, can be characterized as the space

$$D^{s,2}(\mathbb{R}^N) = \left\{ u \in L^{2_s^*}(\mathbb{R}^N) : \int_{\mathbb{R}^N} \int_{\mathbb{R}^N} \frac{|u(x) - u(y)|^2}{|x - y|^{N+2s}} dx dy < +\infty \right\}.$$

$f \in C(\mathbb{R}, \mathbb{R})$ satisfy the following Berestycki–Lions-type condition [19,20]:

(F5) F is not trivial, that is, $F \not\equiv 0$;

(F6)   there exists $C > 0$ such that for every $t \in \mathbb{R}$,

$$|tf(t)| \leq C|t|^{\frac{2N-\mu}{N-2s}};$$

(F7)

$$\lim_{t \to 0} \frac{F(t)}{|t|^2} = \lim_{t \to \infty} \frac{F(t)}{|t|^{\frac{2N-\mu}{N-2s}}} = 0.$$

The second result of this paper is as follows.

Theorem 2. *Suppose that f satisfies (F5)–(F7). Then* (6) *has a nontrivial solution.*

Remark 2. *Notice that the method used in* [13] *is no longer applicable for* (6), *because it relies heavily on the constant potentials. In the zero mass case, we use the perturbation method and the Pohožaev identity established in* [19] *to overcome this difficulty.*

In this article, we make use of the following notation:

- $\|\cdot\|_p$ denotes the usual norm of $L^p(\mathbb{R}^3)$;
- $C, C_i, i = 1, 2, \cdots$, denote various positive constants whose exact values are irrelevant;
- $o(1)$ denotes the infinitesimal as $n \to +\infty$.

2. Ground State Solutions for (1)

Set
$$D^{s,2}(\mathbb{R}^N) := \left\{ u \in L^{2_s^*}(\mathbb{R}^N) : \int_{\mathbb{R}^N} \int_{\mathbb{R}^N} \frac{|u(x) - u(y)|^2}{|x - y|^{N+2s}} dx dy < +\infty \right\},$$

endowed with the Gagliardo (semi)norm

$$[u] := \left(\int_{\mathbb{R}^N} \int_{\mathbb{R}^N} \frac{|u(x) - u(y)|^2}{|x - y|^{N+2s}} dx dy \right)^{1/2}.$$

From [5], we have the following identity:

$$[u]^2 = \int_{\mathbb{R}^N} |(-\Delta)^{\frac{s}{2}} u|^2 dx = \int_{\mathbb{R}^N} |\xi|^{2s} |\mathcal{F}[u](\xi)|^2 d\xi.$$

From [26], $D^{s,2}(\mathbb{R}^N)$ is continuously embedded into $L^{2_s^*}(\mathbb{R}^N)$. Then, we can define the best constant $S > 0$ as

$$S := \sup_{u \in D^{s,2}(\mathbb{R}^N)} \frac{\left(\int_{\mathbb{R}^N} |u|^{2_s^*} dx \right)^{\frac{2}{2_s^*}}}{\int_{\mathbb{R}^N} |(-\Delta)^{\frac{s}{2}} u|^2 dx}.$$

Let

$$E := \left\{ u \in D^{s,2}(\mathbb{R}^N) : \int_{\mathbb{R}^N} V(x)u^2 dx < +\infty \right\}.$$

Under the assumptions (I)–(III), following the idea of ([21], Proposition 2.1) or ([22], Proposition 2.2), we can prove that the Hilbert space E endowed with scalar product and norm

$$(u,v) = \int_{\mathbb{R}^N} [(-\Delta)^{\frac{s}{2}} u (-\Delta)^{\frac{s}{2}} v + V(x)uv]dx, \quad \|u\| = \left(\int_{\mathbb{R}^N} [|(-\Delta)^{\frac{s}{2}} u|^2 + V(x)u^2]dx \right)^{\frac{1}{2}}$$

is compactly embedded into the weighted space $L_K^q(\mathbb{R}^N)$ for every $q \in (2, 2_s^*)$, where $K(x) := [Q(x)]^{2N/(2N-\mu)}$ and

$$L_K^q(\mathbb{R}^N) := \left\{ u : \text{meas}\{u\} < \infty \text{ and } \int_{\mathbb{R}^N} K(x)|u|^q dx < \infty \right\}, \quad \forall q \geq 2.$$

Lemma 1. *Assume that (I)–(III) hold. If (K1) holds, E is compactly embedded in $L_K^q(\mathbb{R}^N)$ for all $q \in (2, 2_s^*)$. If (K2) holds, E is compactly embedded in $L_K^p(\mathbb{R}^N)$.*

Proof. If $(K1)$ holds, then

$$\frac{K(x)}{V(x)} = \frac{Q(x)}{V(x)} [Q(x)]^{\frac{\mu}{2N-\mu}} \in L^\infty(\mathbb{R}^N).$$

Given $\varepsilon > 0$ and fixed $q \in (2, 2_s^*)$, there exist $0 < t_0 < t_1$ and $C > 0$ such that

$$K(x)|t|^q \leq \varepsilon C(V(x)|t|^2 + |t|^{2_s^*}) + CK(x)\chi_{[t_0,t_1]}(|t|)|t|^{2_s^*} \quad \forall t \in \mathbb{R}.$$

Hence,

$$\int_{B_r^c(0)} K(x)|u|^q dx \leq \varepsilon CW(u) + CK(x) \int_{A \cap B_r^c(0)} K(x)dx \quad \forall u \in E, \tag{7}$$

where

$$W(u) = \int_{\mathbb{R}^N} V(x)|u|^2 dx + \int_{\mathbb{R}^N} |u|^{2_s^*} dx$$

and

$$A = \{x \in \mathbb{R}^N : s_0 \leq |u(x)| \leq s_1\}.$$

Let $\{v_n\}$ be a sequence such that $v_n \rightharpoonup v$ in E, then there exists a constant $M_1 > 0$ such that

$$\int_{\mathbb{R}^N} [|(-\Delta)^{\frac{s}{2}} v_n|^2 + V(x)|v_n|^2]dx \leq M_1 \text{ and } \int_{\mathbb{R}^N} |v_n|^{2_s^*} dx \leq M_1 \quad \forall n \in \mathbb{N},$$

which implies that $\{W(v_n)\}$ is bounded. On the other hand, setting

$$A_n = \{x \in \mathbb{R}^N : s_0 \leq |v_n(x)| \leq s_1\},$$

we have

$$s_0^{2_s^*}|A_n| \leq \int_{A_n} |v_n|^{2_s^*} dx \leq M_1 \quad \forall n \in \mathbb{N}$$

and so $\sup_{n \in \mathbb{N}} |A_n| < +\infty$. Therefore, from (II), there is $r > 0$ such that

$$\int_{A_n \cap B_r^c(0)} K(x)dx < \frac{\varepsilon}{s_1^{2_s^*}} \quad \forall n \in \mathbb{N}. \tag{8}$$

Combining (7) and (8), we have

$$\int_{B_r^c(0)} K(x)|v_n|^q dx < \varepsilon C M_1 + s_1^{2_s^*} \int_{F_n \cap B_r^c(0)} K(x)dx < (CM_1 + 1)\varepsilon \ \ \forall n \in \mathbb{N}. \tag{9}$$

By $q \in (2, 2_s^*)$, we have from Sobolev embeddings that

$$\lim_{n \to +\infty} \int_{B_r(0)} K(x)|v_n|^q dx = \int_{B_r(0)} K(x)|v|^q dx. \tag{10}$$

Combining (9) and (10), we have

$$\lim_{n \to +\infty} \int_{\mathbb{R}^N} K(x)|v_n|^q dx = \int_{\mathbb{R}^N} K(x)|v|^q dx,$$

which yields

$$v_n \to v \text{ in } L_K^q(\mathbb{R}^N) \ \ \forall q \in (2, 2_s^*).$$

Next, we suppose that $(K2)$ holds. For each $x \in \mathbb{R}^N$ fixed, we observe that the function

$$g(t) = V(x)t^{2-p} + t^{2_s^*-p} \ \ \forall t > 0$$

has $C_p V(x)^{\frac{2_s^*-p}{2_s^*-2}}$ as its minimum value, where

$$C_p = \left(\frac{p-2}{2_s^* - p} \right)^{\frac{2-p}{2_s^*-2}} + \left(\frac{p-2}{2_s^* - p} \right)^{\frac{2_s^*-p}{2_s^*-2}}.$$

Hence

$$C_p V(x)^{\frac{2_s^*-p}{2_s^*-2}} \leq V(x)t^{2-p} + t^{2_s^*-p} \ \ \forall x \in \mathbb{R}^N \text{ and } t > 0.$$

Combining this inequality with $(K2)$, given $\varepsilon \in (0, C_p)$, there exists $r > 0$ large enough such that

$$K(x)|t|^p \leq \varepsilon(V(x)|t|^2 + |t|^{2_s^*}) \ \ \forall t \in \mathbb{R} \text{ and } |x| \geq r,$$

leading to

$$\int_{B_r^c(0)} K(x)|u|^p dx \leq \varepsilon \int_{B_r^c(0)} (V(x)|u|^2 + |u|^{2_s^*})dx \ \ \forall u \in E.$$

Let $\{v_n\}$ be a sequence such that $v_n \rightharpoonup v$ in E, then there exists a constant $M_2 > 0$ such that

$$\int_{\mathbb{R}^N} V(x)|v_n|^2 dx \leq M_2 \text{ and } \int_{\mathbb{R}^N} |v_n|^{2_s^*} dx \leq M_2 \ \ \forall n \in \mathbb{N},$$

and so,

$$\int_{B_r^c(0)} K(x)|v_n|^p dx \leq 2\varepsilon M_2 \ \ \forall n \in \mathbb{N}. \tag{11}$$

Since $p \in (2, 2_s^*)$ and K is a continuous function, we have

$$\lim_{n \to +\infty} \int_{B_r^c(0)} K(x)|v_n|^p dx = \int_{B_r^c(0)} K(x)|v|^p dx. \tag{12}$$

From (11) and (12), we have

$$\lim_{n \to +\infty} \int_{\mathbb{R}^N} K(x)|v_n|^p dx = \int_{\mathbb{R}^N} K(x)|v|^p dx.$$

Therefore

$$v_n \to v \text{ in } L_K^p(\mathbb{R}^N).$$

$\square$

Lemma 2. *(Hardy–Littlewood–Sobolev inequality, see [26]). Let $1 < r, t < \infty$, and $\mu \in (0, N)$ with $\frac{1}{r} + \frac{1}{t} = 2 - \frac{\mu}{N}$. If $\phi \in L^r(\mathbb{R}^N)$ and $\psi \in L^t(\mathbb{R}^N)$, then there exists a constant $C(N, \mu, r, t) > 0$, such that*

$$\int_{\mathbb{R}^N} \int_{\mathbb{R}^N} \frac{\phi(x)\psi(y)}{|x-y|^\mu} dxdy \leq C(N, \mu, r, t)\|\phi\|_r\|\psi\|_t.$$

Lemma 3. *Assume that (I)–(III) and (F1)–(F3) hold. Then for $u \in E$*

$$\left| \int_{\mathbb{R}^N} \int_{\mathbb{R}^N} \frac{Q(x)Q(y)F(u(x))F(u(y))}{|x-y|^\mu} dxdy \right| < +\infty, \tag{13}$$

and there exists a constant $C_1 > 0$ such that

$$\left| \int_{\mathbb{R}^N} \int_{\mathbb{R}^N} \frac{Q(x)Q(y)F(u(x))f(u(y))v(y)}{|x-y|^\mu} dxdy \right| < C_1\|v\|, \ \forall v \in E. \tag{14}$$

Furthermore, let $\{u_n\} \subset E$ be a sequence such that $u_n \rightharpoonup u$ in E, then

$$\lim_{n \to \infty} \int_{\mathbb{R}^N} \int_{\mathbb{R}^N} \frac{Q(x)Q(y)[F(u_n(x))F(u_n(y)) - F(u(x))F(u(y))]}{|x-y|^\mu} dxdy = 0 \tag{15}$$

and

$$\lim_{n \to \infty} \int_{\mathbb{R}^N} \int_{\mathbb{R}^N} \frac{Q(x)Q(y)F(u_n(x))f(u_n(y))[u_n(y) - u(y)]}{|x-y|^\mu} dxdy = 0. \tag{16}$$

Proof. Set

$$\beta = \begin{cases} 2, & \text{if } (K1) \text{ holds}, \\ p, & \text{if } (K2) \text{ holds}. \end{cases}$$

By $(F1), (F2)$, Lemma 2, Hölder inequality and Sobolev inequality, we have

$$\begin{aligned}
\int_{\mathbb{R}^N} K(x)|F(u)|^{\frac{2N}{2N-\mu}} dx &\leq C_1 \int_{\mathbb{R}^N} K(x) \left[|u|^{\frac{\beta(2N-\mu)}{2N}} + |u|^{\frac{2N-\mu}{N-2s}} \right]^{\frac{2N}{2N-\mu}} dx \\
&\leq C_2 \int_{\mathbb{R}^N} K(x)|u|^\beta dx + C_2 \int_{\mathbb{R}^N} |u|^{2_s^*} dx \\
&\leq C_3(\|u\|^\beta + [u]^{2_s^*}), \ \forall u \in E
\end{aligned} \tag{17}$$

and

$$\int_{\mathbb{R}^N} K(x)|f(u)v|^{\frac{2N}{2N-\mu}}dx \le C_1 \int_{\mathbb{R}^N} K(x)\left[|u|^{\frac{\beta(2N-\mu)-2N}{2N}} + |u|^{\frac{N-\mu+2s}{N-2s}}\right]^{\frac{2N}{2N-\mu}} |v|^{\frac{2N}{2N-\mu}}dx$$
$$\le C_4 \int_{\mathbb{R}^N} [K(x)]^{\frac{\beta(2N-\mu)-2N}{\beta(2N-\mu)}}|u|^{\frac{\beta(2N-\mu)-2N}{2N-\mu}} [K(x)]^{\frac{2N}{\beta(2N-\mu)}}|v|^{\frac{2N}{2N-\mu}}dx$$
$$+ C_5 \int_{\mathbb{R}^N} |u|^{\frac{2N(N+2s-\mu)}{(N-2s)(2N-\mu)}}|v|^{\frac{2N}{2N-\mu}}dx \tag{18}$$
$$\le C_6 \left[\|u\|^{\frac{\beta(2N-\mu)-2N}{2N-\mu}} + \|u\|^{\frac{2N(N+2s-\mu)}{(N-2s)(2N-\mu)}}\right]\|v\|^{\frac{2N}{2N-\mu}}, \quad \forall u, v \in E.$$

Applying Lemma 2 and (17), we have

$$\left|\int_{\mathbb{R}^N}\int_{\mathbb{R}^N} \frac{Q(x)Q(y)F(u(x))\tilde{F}(u(y))}{|x-y|^\mu}dxdy\right|$$
$$\le C_7 \left[\int_{\mathbb{R}^N} K(x)|F(u)|^{\frac{2N}{2N-\mu}}dx\right]^{\frac{2N-\mu}{N}} \tag{19}$$
$$\le C_8 \left[\|u\|^{\frac{\beta(2N-\mu)}{N}} + \|u\|^{\frac{2(2N-\mu)}{N-2s}}\right], \quad \forall u \in E,$$

which yields (13) holds. Similarly, we have

$$\left|\int_{\mathbb{R}^N}\int_{\mathbb{R}^N} \frac{Q(x)Q(y)F(u(x))f(u(y))v(y)}{|x-y|^\mu}dxdy\right|$$
$$\le C_9 \left[\int_{\mathbb{R}^N} K(x)|F(u)|^{\frac{2N}{2N-\mu}}dx\right]^{\frac{2N-\mu}{2N}} \left[\int_{\mathbb{R}^N} K(x)|f(u)v|^{\frac{2N}{2N-\mu}}dx\right]^{\frac{2N-\mu}{2N}}, \quad \forall u, v \in E, \tag{20}$$

which, together with (17) and (18), implies that (14) holds.

Similar to ([21], Lemma 2), by (F_2), (F_3), and Lemma 2, we have

$$\lim_{n\to\infty} \int_{\mathbb{R}^N} K(x)|F(u_n) - F(u)|^{\frac{2N}{2N-\mu}}dx = 0, \quad \lim_{n\to\infty} \int_{\mathbb{R}^N} K(x)|f(u_n)|^{\frac{2N}{2N-\mu}}|u_n - u|^{\frac{2N}{2N-\mu}}dx = 0. \tag{21}$$

Combining (18), (20), and (21), we deduce that (15) and (16) hold. $\square$

The energy functional $\Phi : E \mapsto \mathbb{R}$ given by

$$\Phi(u) := \frac{1}{2}\int_{\mathbb{R}^N} |(-\Delta)^{\frac{s}{2}}u|^2 dx + \frac{1}{2}\int_{\mathbb{R}^N} V(x)|u|^2 dx - \frac{1}{2}\int_{\mathbb{R}^N}\int_{\mathbb{R}^N} \frac{Q(x)Q(y)F(u(x))F(u(y))}{|x-y|^\mu}dxdy. \tag{22}$$

By Lemmas 2 and 3, Φ is well-defined and belongs to C^1-class. Moreover, we have

$$\langle\Phi'(u), v\rangle = \int_{\mathbb{R}^N} (-\Delta)^{\frac{s}{2}}u(-\Delta)^{\frac{s}{2}}v dx + \int_{\mathbb{R}^N} V(x)uv dx$$
$$- \int_{\mathbb{R}^N}\int_{\mathbb{R}^N} \frac{Q(x)Q(y)F(u(x))f(u(y))v(y)}{|x-y|^\mu}dxdy, \quad \forall u, v \in E. \tag{23}$$

Lemma 4. *Assume that* (F1)–(F3) *hold. Then, for all* $t \ge 0$ *and* $\tau_1, \tau_2 \in \mathbb{R}$,

$$l(t, \tau_1, \tau_2) := F(t\tau_1)F(t\tau_2) - F(\tau_1)F(\tau_2) + \frac{1-t^2}{2}[F(\tau_1)f(\tau_2)\tau_2 + F(\tau_2)f(\tau_1)\tau_1] \ge 0. \tag{24}$$

Proof. Firstly, it follows from $(F1)$ that $f(0) = 0$. By $(F3)$, we have

$$f(\tau) \geq 0, \ \forall \tau \geq 0; \quad f(\tau) \leq 0, \ \forall \tau \leq 0; \quad F(\tau) \geq 0, \ \forall \tau \in \mathbb{R}$$

and

$$f(\tau)\tau \geq \int_0^\tau f(t)dt = F(\tau), \quad \forall \tau \in \mathbb{R}. \tag{25}$$

It is easy to verify that (24) holds for $t = 0$. For $\tau \neq 0$, we have from (25) that

$$\left[\frac{F(\tau)}{\tau}\right]' = \frac{f(\tau)\tau - F(\tau)}{\tau^2} \geq 0. \tag{26}$$

For every $\tau_1, \tau_2 \in \mathbb{R}$, we deduce from $(F3)$ and (26) that

$$
\begin{aligned}
&\frac{d}{dt} l(t, \tau_1, \tau_2) \\
&= \tau_1 \tau_2 t \left[\frac{F(t\tau_1)}{t\tau_1} f(t\tau_2) + \frac{F(t\tau_2)}{t\tau_2} f(t\tau_1) - \frac{F(\tau_1)}{\tau_1} f(\tau_2) - \frac{F(\tau_2)}{\tau_2} f(\tau_1)\right] \\
&\begin{cases} \geq 0, \ t \geq 1, \\ \leq 0, \ 0 < t < 1, \end{cases}
\end{aligned}
$$

which implies that $l(t, \tau_1, \tau_2) \geq l(1, \tau_1, \tau_2) = 0$ for all $t > 0$ and $\tau_1, \tau_2 \in \mathbb{R}$. $\square$

Lemma 5. *Assume that* (I)–(III) *and* $(F1)$–$(F4)$ *hold. Then*

$$\Phi(u) \geq \Phi(tu) + \frac{1 - t^2}{2} \langle \Phi'(u), u \rangle, \ \forall u \in E, \ t \geq 0. \tag{27}$$

Proof. By (22), (23), and (24), we have

$$
\begin{aligned}
&\Phi(u) - \Phi(tu) - \frac{1 - t^2}{2} \langle \Phi'(u), u \rangle \\
&= \frac{1}{2} \int_{\mathbb{R}^N} \int_{\mathbb{R}^N} \frac{1}{|x - y|^\mu} \Big[F(tu(x))F(tu(y)) - F(u(x))F(u(y)) \\
&\qquad + \frac{1 - t^2}{2} \big(F(u(x))f(u(y))u(y) + F(u(y))f(u(x))u(x) \big) \Big] dxdy \\
&= \frac{1}{2} \int_{\mathbb{R}^N} \int_{\mathbb{R}^N} \frac{l(t, u(x), u(y))}{|x - y|^\mu} dxdy \\
&\geq 0, \ \forall u \in E, \ t \geq 0.
\end{aligned}
$$

$\square$

Corollary 1. *Assume that* (I)–(III) *and* $(F1)$–$(F4)$ *hold. Let*

$$\mathcal{N} := \{u \in E \setminus \{0\} : \langle \Phi'(u), u \rangle = 0\}.$$

Then

$$\Phi(u) = \max_{t \geq 0} \Phi(tu), \ \forall u \in \mathcal{N}.$$

Lemma 6. *Assume that* (I)–(III) *and* $(F1)$–$(F4)$ *hold. Then, for any* $u \in E \setminus \{0\}$*, there exists* $t_u > 0$ *such that* $t_u u \in \mathcal{N}$*.*

Proof. Let $u \in E \setminus \{0\}$ be fixed. Define a function $\zeta(t) := \Phi(tu)$ on $(0, \infty)$. By (22) and (23), we have

$$\zeta'(t) = 0 \iff t\|u\|^2 - \int_{\mathbb{R}^N} \int_{\mathbb{R}^N} \frac{Q(x)Q(y)F(tu(x))F(tu(y)))f(tu(y))u(y)}{|x-y|^\mu}\,dxdy = 0$$
$$\iff tu \in \mathcal{N}.$$

By (19), we have for $u \in E$

$$\Phi(u) \geq \begin{cases} \frac{1}{2}\|u\| - C_8\left[\|u\|^{\frac{4N-2\mu}{N}} + \|u\|^{\frac{4N-2\mu}{N-2s}}\right], & \text{if } (K1) \text{ holds,} \\[2mm] \frac{1}{2}\|u\| - C_8\left[\|u\|^{\frac{2pN-p\mu}{N}} + \|u\|^{\frac{4N-2\mu}{N-2s}}\right], & \text{if } (K2) \text{ holds,} \end{cases} \tag{28}$$

which implies that there exists $\rho_0 > 0$ such that

$$\delta_0 := \inf_{\|u\|=\rho_0} \Phi(u) > 0. \tag{29}$$

Therefore, $\lim_{t\to 0} \zeta(t) = 0$ and $\zeta(t) > 0$ for small $t > 0$. By $(F4)$, for t large, we have

$$\zeta(t) = \frac{t^2}{2}\left[\|u\|^2 - \frac{1}{2}\int_{\mathbb{R}^N}\int_{\mathbb{R}^N} \frac{Q(x)F(tu(x))}{|tu(x)|}\,\frac{Q(y)F(tu(y))}{|tu(y)|}\,\frac{|u(x)u(y)|}{|x-y|^\mu}\,dxdy\right] < 0. \tag{30}$$

Therefore $\max_{t\in[0,\infty)} \zeta(t)$ is achieved at some $t_u > 0$ so that $\zeta'(t_u) = 0$ and $t_u u \in \mathcal{N}$. $\quad\square$

Lemma 7. *Assume that* (I)–(III) *and* $(F1)$–$(F4)$ *hold. Then*

$$\inf_{u\in\mathcal{N}} \Phi(u) := c = \inf_{u\in E\setminus\{0\}} \max_{t\geq 0} \Phi(tu) > 0.$$

Proof. Corollary 1 and Lemma 6 imply that

$$c = \inf_{u\in E\setminus\{0\}} \max_{t\geq 0} \Phi(tu).$$

By (22) and (29),

$$c \geq \inf_{u\in E\setminus\{0\}} \Phi\left(\frac{\rho_0}{\|u\|}u\right) = \inf_{\|u\|=\rho_0} \Phi(u) > 0.$$

$\square$

Next, we will seek a Cerami sequence for Φ outside $\mathcal{N}$ by using the diagonal method, which is used in [25,27,28].

Lemma 8. *Assume that* (I)–(III) *and* $(F1)$–$(F4)$ *hold. Then there exist* $\{u_n\} \subset E$ *and* $c^* \in (0, c]$ *such that*

$$\Phi(u_n) \to c^*, \quad (1 + \|u_n\|)\|\Phi'(u_n)\| \to 0, \tag{31}$$

as $n \to \infty$*.*

Proof. For $c = \inf\limits_{\mathcal{N}} \Phi$, we can choose a sequence $\{v_k\} \subset \mathcal{N}$ such that

$$c \leq \Phi(v_k) < c + \frac{1}{k}, \quad k \in \mathbb{N}. \tag{32}$$

By (29) and (30), it is easy to verify that $\Phi(0) = 0$, $\Phi(Tv_k) < 0$ when T is large enough, and $\Phi(u) \geq \delta_0 > 0$ when $\|u\| = \rho_0$. Therefore, from Mountain Pass Lemma ([29]), there is a sequence $\{u_{n,k}\}$ such that

$$\Phi(u_{k,n}) \to c_k \in [\delta_0, \sup_{t \in [0,1]} \Phi(tv_k)], \quad (1 + \|u_{k,n}\|)\|\Phi'(u_{k,n})\| \to 0, \quad k \in \mathbb{N}. \tag{33}$$

By Corollary 1 and $\{v_k\} \subset \mathcal{N}$, we have

$$\Phi(tv_k) \leq \Phi(v_k), \quad \forall\, t \geq 0. \tag{34}$$

It follows from (34) that $\Phi(v_k) = \sup_{t \in [0,1]} \Phi(tv_k)$. Hence, by (32)–(34), we have

$$\Phi(w_{k,n}) \to c_k \in \left[\delta_0, c + \frac{1}{k}\right), \quad (1 + \|u_{k,n}\|)\|\Phi'(u_{k,n})\| \to 0, \quad k \in \mathbb{N}.$$

Then, we can choose $\{n_k\} \subset \mathbb{N}$ such that

$$\Phi(u_{k,n_k}) \in \left[\delta_0, c + \frac{1}{k}\right), \quad (1 + \|u_{k,n_k}\|)\|\Phi'(u_{k,n_k})\| < \frac{1}{k}, \quad k \in \mathbb{N}.$$

Let $u_k = u_{k,n_k}$, $k \in \mathbb{N}$. Therefore, up to a subsequence, we have

$$\Phi(u_n) \to c^* \in [\delta_0, c], \quad (1 + \|u_n\|)\|\Phi'(u_n)\| \to 0.$$

$\square$

Lemma 9. *Assume that (I)–(III) and (F1)–(F4) hold. Then, the sequence $\{u_n\}$ satisfying (31) is bounded in E.*

Proof. Arguing by contradiction, suppose that $\|u_n\| \to \infty$. Let $v_n = \frac{u_n}{\|u_n\|}$, then $\|v_n\| = 1$. Passing to a subsequence, we have $v_n \rightharpoonup v$ in E. There are two possible cases: (i). $v = 0$; (ii) $v \neq 0$.

Case (i) $v = 0$. In this case

$$\left| \int_{\mathbb{R}^N} \frac{Q(x)Q(y)F(2\sqrt{c^* + 1}v_n(x))F(2\sqrt{c^* + 1}v_n(y))}{|x - y|^\mu} dxdy \right|$$

$$\leq C_1 \left[\int_{\mathbb{R}^N} K(x)|F(2\sqrt{c^* + 1}v_n(x))|^{\frac{2N}{2N-\mu}} dx \right]^{\frac{2N-\mu}{N}} \tag{35}$$

$$= o(1).$$

Combining (27), (31), and (35), we have

$$c^* + o(1) = \Phi(u_n)$$

$$\geq \Phi\left(\frac{2\sqrt{c^*+1}}{\|u_n\|}u_n\right) + \frac{1 - \left(\frac{2\sqrt{c^*+1}}{\|u_n\|}\right)^2}{2}\langle\Phi'(u_n), u_n\rangle$$

$$= \Phi(2\sqrt{c^*+1}\,v_n) + o(1)$$

$$= 2(c^*+1) + o(1),$$

which is a contradiction.

Case (ii) $v \neq 0$. In this case, since $|u_n| = |v_n|\|u_n\|$ and $u_n/\|u_n\| \to v$ a.e. in $\mathbb{R}^N$, we have $\lim\limits_{n\to\infty} |u_n(x)| = \infty$ for $x \in \{y \in \mathbb{R}^N : v(x) \neq 0\}$. Hence, it follows from (22), (31), (F4), and Fatou's lemma that

$$0 = \lim_{n\to\infty} \frac{c^* + o(1)}{\|u_n\|^2} = \lim_{n\to\infty} \frac{\Phi(u_n)}{\|u_n\|^2}$$

$$= \frac{1}{2} - \frac{1}{2}\lim_{n\to\infty}\int_{\mathbb{R}^N}\int_{\mathbb{R}^N}\frac{Q(x)F(u_n(x))}{|u_n(x)|}\frac{Q(y)F(u_n(y))}{|u_n(y)|}\frac{|v_n(x)v_n(y)|}{|x-y|^\mu}dxdy$$

$$\leq \frac{1}{2} - \frac{1}{2}\int_{\mathbb{R}^N}\int_{\mathbb{R}^N}\liminf_{n\to\infty}\frac{Q(x+k_n)F(u_n(x))}{|u_n(x)|}\frac{Q(y+k_n)F(u_n(y))}{|u_n(y)|}\frac{|v_n(x)v_n(y)|}{|x-y|^\mu}dxdy$$

$$= -\infty.$$

This contradiction shows that $\{u_n\}$ is bounded in E. $\square$

Proof of Theorem 1. In view of Lemmas 8 and 9, there exists a bounded sequence $\{u_n\} \subset E$ such that (31) holds. Passing to a subsequence, we have $u_n \rightharpoonup u$ in E. Thus, it follows from (22), (23), (31), and Lemma 3 that

$$\|u_n - u\|^2 = \langle\Phi'(u_n), u_n - u\rangle + \int_{\mathbb{R}^N}\int_{\mathbb{R}^N}\frac{Q(x)Q(y)F(u_n(x))f(u_n(y))[u_n(y) - u(y)]}{|x-y|^\mu} = o(1),$$

which implies that $\Phi'(u) = 0$ and $\Phi(u) = c^* \in (0, c]$. Moreover, since $u \in \mathcal{N}$, we have $\Phi(u) \geq c$. Hence, $u \in E$ is a ground state solution for (1) with $\Phi(u) = c > 0$. $\square$

3. Zero Mass Case

In this section, we consider the zero mass case, and give the proof of Theorem 2. In the following, we suppose that (F5)–(F7) and $\mu < 4s$ hold. Fix $q \in (2, \frac{2N-\mu}{N-2s})$, by (F7), for every $\epsilon > 0$ there is $C_\epsilon > 0$ such that

$$|f(t)t| \leq \epsilon(|t|^2 + |t|^{\frac{2N-\mu}{N-2s}}) + C_\epsilon|t|^q, \quad |F(t)| \leq \epsilon(|t|^2 + |t|^{\frac{2N-\mu}{N-2s}}) + C_\epsilon|t|^q, \quad \forall t \in \mathbb{R}. \tag{36}$$

To find nontrivial solutions for (6), we study the approximating problem

$$\begin{cases} (-\Delta)^s u + \varepsilon u = \left(\frac{1}{|x|^\mu} * F(u)\right) f(u), & \text{in } \mathbb{R}^N, \\ u \in H^s(\mathbb{R}^N), \end{cases} \tag{37}$$

where $\varepsilon \geq 0$ is a small parameter. The energy functional associated to (37) is

$$\Phi_\varepsilon(u) = \frac{1}{2}\int_{\mathbb{R}^N}[|(-\Delta)^{\frac{s}{2}}u|^2 + \varepsilon u^2]dx - \frac{1}{2}\int_{\mathbb{R}^N}\int_{\mathbb{R}^N}\frac{F(u(x))F(u(y))}{|x-y|^\mu}dxdy. \tag{38}$$

By using $(F5)$–$(F7)$ and Lemma 2, it is easy to check that $\Phi_0 \in C^1(D^{s,2}(\mathbb{R}^N), \mathbb{R})$ and $\Phi_\varepsilon \in C_1(H^s(\mathbb{R}^N), \mathbb{R})$ for every $\varepsilon > 0$. Moreover, for every $\varepsilon \geq 0$,

$$\langle \Phi'_\varepsilon(u), v \rangle = \int_{\mathbb{R}^N}[(-\Delta)^{\frac{s}{2}}u(-\Delta)^{\frac{s}{2}}v + \varepsilon uv]dx - \frac{1}{2}\int_{\mathbb{R}^N}\int_{\mathbb{R}^N}\frac{F(u(x))f(u(y))v(y)}{|x-y|^\mu}dxdy. \tag{39}$$

In view of ([19], Proposition 2), for every $\varepsilon > 0$, any critical point u of Φ_ε in $H^s(\mathbb{R}^N)$ satisfies the following Pohožaev identity

$$\mathcal{P}_\varepsilon(u) := \frac{N-2s}{2}\int_{\mathbb{R}^N}|(-\Delta)^{\frac{s}{2}}u|^2dx + \frac{N}{2}\varepsilon\int_{\mathbb{R}^N}|u|^2dx - \frac{2N-\mu}{2}\int_{\mathbb{R}^N}\int_{\mathbb{R}^N}\frac{F(u(x))F(u(y))}{|x-y|^\mu}dxdy \tag{40}$$
$$= 0.$$

For every $\varepsilon > 0$, let

$$\mathcal{M}_\varepsilon := \{u \in H^s(\mathbb{R}^N) \setminus \{0\} : \Phi'_\varepsilon(u) = 0\},$$
$$\Gamma_\varepsilon := \{\gamma \in C([0,1], H^s(\mathbb{R}^N)) : \gamma(0) = 0, \Phi_\varepsilon(\gamma(1)) < 0\},$$
$$c_\varepsilon := \inf_{\gamma \in \Gamma_\varepsilon}\max_{t \in [0,1]}\Phi_\varepsilon(\gamma(t)).$$

Lemma 10. *For every $\varepsilon > 0$, (37) has a ground state solution $u_\varepsilon \in H^s(\mathbb{R}^N)$ such that $0 < \Phi_\varepsilon(u_\varepsilon) = \inf_{\mathcal{M}_\varepsilon}\Phi_\varepsilon = c_\varepsilon$. Moreover, there exists a constant $K_0 > 0$ independent of ε such that $c_\varepsilon \leq K_0$ for all $\varepsilon \in (0,1]$.*

Proof. In view of ([19], Theorem 1.3), under the assumption $(F5)$–$(F7)$, for every $\varepsilon > 0$, (37) has a ground state solution $u_\varepsilon \in H^s(\mathbb{R}^N)$ such that $0 < \Phi_\varepsilon(u_\varepsilon) = \inf_{\mathcal{M}_\varepsilon}\Phi_\varepsilon = c_\varepsilon$. Let $\gamma \in \Gamma_1$, since $\Phi_\varepsilon(u) \leq \Phi_1(u)$ for $u \in H^s(\mathbb{R}^N)$ and $\varepsilon \in (0,1]$, we have $\gamma \in \Gamma_\varepsilon$ for $\varepsilon \in (0,1]$, and so

$$c_\varepsilon \leq \max_{t \in [0,1]}\Phi_\varepsilon(\gamma(t)) = \Phi_\varepsilon(\gamma(t_\varepsilon)) \leq \Phi_1(\gamma(t_\varepsilon)) \leq \max_{t \in [0,1]}\Phi_1(\gamma(t)) := K_0, \quad \forall \varepsilon \in (0,1],$$

where $t_\varepsilon \in (0,1)$. $\square$

Lemma 11. *There exists a constant $K_1 > 0$ independent of ε such that*

$$[u_\varepsilon] \geq K_1, \quad \forall u_\varepsilon \in \mathcal{M}_\varepsilon. \tag{41}$$

Proof. Since $\langle \Phi_\varepsilon'(u_\varepsilon), u_\varepsilon \rangle = 0$ for $u_\varepsilon \in \mathcal{M}_\varepsilon$, from (F6), (39), and Sobolev inequality, we have

$$
\begin{aligned}
[u_\varepsilon]^2 &= \int_{\mathbb{R}^N} |(-\Delta)^{\frac{s}{2}} u_\varepsilon|^2 dx \le \int_{\mathbb{R}^N} [|(-\Delta)^{\frac{s}{2}} u_\varepsilon|^2 + \varepsilon u_\varepsilon^2] dx \\
&= \int_{\mathbb{R}^N} \int_{\mathbb{R}^N} \frac{F(u_\varepsilon(x)) f(u_\varepsilon(y)) u_\varepsilon(y)}{|x-y|^\mu} dx dy \\
&\le C_1 \left(\int_{\mathbb{R}^N} |F(u_\varepsilon)|^{\frac{2N}{2N-\mu}} dx \right)^{\frac{2N-\mu}{2N}} \left(\int_{\mathbb{R}^N} |f(u_\varepsilon) u_\varepsilon|^{\frac{2N}{2N-\mu}} dx \right)^{\frac{2N-\mu}{2N}} \\
&\le C_2 \left(\int_{\mathbb{R}^N} |u_\varepsilon|^{\frac{2N}{N-2s}} dx \right)^{\frac{2N-\mu}{N}} \\
&\le C_2 S^{\frac{2N-\mu}{N-2s}} [u_\varepsilon]^{\frac{2(2N-\mu)}{N-2s}}, \quad \forall u_\varepsilon \in \mathcal{M}_\varepsilon,
\end{aligned}
$$

which, together with $(2N - \mu)/(N - 2s) > 1$, implies that (41) holds. $\square$

The following lemma is a version of Lions' concentration-compactness Lemma for fractional Laplacian.

Lemma 12. *([18]) Assume $\{u_n\}$ is a bounded sequence in $H^s(\mathbb{R}^N)$, which satisfies*

$$
\lim_{\substack{n \to +\infty \\ y \in \mathbb{R}^N}} \sup \int_{B_1(y)} |u_n(x)|^2 dx = 0.
$$

Then $u_n \to 0$ in $L^q(\mathbb{R}^N)$ for $q \in (2, 2_s^*)$.

Proof of Theorem 2. We choose a sequence $\{\varepsilon_n\} \subset (0, 1]$ such that $\varepsilon_n \searrow 0$. In view of Lemma 10, there exists a sequence $\{u_{\varepsilon_n}\} \subset \mathcal{M}_{\varepsilon_n}$ such that $0 < \Phi_{\varepsilon_n}(u_{\varepsilon_n}) = \inf_{\mathcal{M}_{\varepsilon_n}} \Phi_{\varepsilon_n} = c_{\varepsilon_n} \le K_0$. For simplicity, we use u_n instead of u_{ε_n}. Now, we prove that $\{u_n\}$ is bounded in $D^{s,2}(\mathbb{R}^N)$. Since $\mathcal{P}_{\varepsilon_n}(u_n) = 0$ for $u_n \in \mathcal{M}_{\varepsilon_n}$, it follows from (38) and (40) that

$$
\begin{aligned}
K_0 \ge c_{\varepsilon_n} &= \Phi_{\varepsilon_n}(u_n) - \frac{1}{2N - \mu} \mathcal{P}_{\varepsilon_n}(u_n) \\
&= \left[\frac{1}{2} - \frac{N - 2s}{2(2N - \mu)} \right] [u_n]^2 + \left[\frac{1}{2} - \frac{N}{2(2N - \mu)} \right] \varepsilon_n \|u_n\|_2^2.
\end{aligned}
\tag{42}
$$

Thus, $\{u_n\}$ is bounded in $D^{s,2}(\mathbb{R}^N)$ and $L^2(\mathbb{R}^N)$. If

$$
\delta := \lim_{\substack{n \to \infty \\ y \in \mathbb{R}^N}} \sup \int_{B_1(y)} |u_n|^2 dx = 0.
$$

Then, by Lemma 12, for $q \in (2, \frac{2N-\mu}{N-2s})$, we have

$$
\int_{\mathbb{R}^N} |u_n|^{\frac{4N}{2N-\mu}} dx \to 0, \quad \int_{\mathbb{R}^N} |u_n|^{\frac{2Nq}{2N-\mu}} dx \to 0.
$$

Therefore, by (36) and Sobolev embedding for $D^{s,2}(\mathbb{R}^N)$, for every $\epsilon > 0$ there exists $C_\epsilon > 0$ such that

$$\left| \int_{\mathbb{R}^N} |F(u_n)|^{\frac{2N}{2N-\mu}} dx \right| \leq \epsilon \left[\int_{\mathbb{R}^N} \left(|u_n|^{\frac{4N}{2N-\mu}} + |u_n|^{2_s^*} \right) dx \right] + C_\epsilon \int_{\mathbb{R}^N} |u_n|^{\frac{2Nq}{2N-\mu}} dx$$
$$\leq \epsilon C + o(1).$$

By the arbitrariness of ϵ, we get

$$\int_{\mathbb{R}^N} |F(u_n)|^{\frac{2N}{2N-\mu}} dx \to 0. \tag{43}$$

Combining (36), (43), and Lemma 2, we have

$$\left| \int_{\mathbb{R}^N} \int_{\mathbb{R}^N} \frac{F(u_n(x)) f(u_n(y)) u_n(y)}{|x-y|^\mu} dx dy \right|$$
$$\leq C_1 \left(\int_{\mathbb{R}^N} |F(u_n)|^{\frac{2N}{2N-\mu}} dx \right)^{\frac{2N-\mu}{2N}} \left(\int_{\mathbb{R}^N} |f(u_n) u_n|^{\frac{2N}{2N-\mu}} dx \right)^{\frac{2N-\mu}{2N}} \tag{44}$$
$$= o(1).$$

Notice that $\{u_n\}$ is bounded in $L^2(\mathbb{R}^N)$, we have from (44) and $u_n \in \mathcal{M}_{\varepsilon_n}$ that $[u_n]^2 = o(1)$. This contradicts (41). Thus, we get $\delta > 0$. Passing to a subsequence, there exists a sequence $\{y_n\} \subset \mathbb{R}^N$ such that

$$\int_{B_{1+\sqrt{N}}(y_n)} |u_n|^2 dx > \frac{\delta}{2}.$$

Let $\tilde{u}_n(x) = u_n(x + y_n)$. Then

$$\Phi'_{\varepsilon_n}(\tilde{u}_n) = 0, \quad \Phi_{\varepsilon_n}(\tilde{u}_n) = c_{\varepsilon_n}$$

and

$$\int_{B_{1+\sqrt{N}}(0)} |\tilde{u}_n|^2 dx > \frac{\delta}{2}. \tag{45}$$

Passing to a subsequence, we have $\tilde{u}_n \rightharpoonup u_0$ in $D^{s,2}(\mathbb{R}^N)$. Clearly, (45) implies that $u_0 \neq 0$. By the standard argument, $u_0 \in D^{s,2}(\mathbb{R}^N)$ is a nontrivial solution for (6). $\square$

4. Conclusions

In this work, we study a class of nonlinear Choquard equation driven by the fractional Laplacian. When potential function vanishes at infinity and the Nehari-type monotonicity condition for the nonlinearity is not satisfied, we prove that the fractional Choquard equation has a ground state solution by using the non-Nehari manifold method. Unlike the Nehari manifold method, the main idea of our approach lies in finding a minimizing sequence for the energy functional outside the Nehari manifold by using the diagonal method. Moreover, by using a perturbation method, we obtain a nontrivial solution in the zero mass case.

Author Contributions: All authors contributed equally in writing this article. All authors read and approved the final manuscript.

Funding: This work is supported by Hainan Natural Science Foundation (Grant No.118MS002 and No.117005), the National Natural Science Foundation of China (Grant No.11861028 and No.11461016), Young Foundation of Hainan University (Grant No.hdkyxj201718).

Conflicts of Interest: The authors declare no conflict of interest.

References

1. Bertoin, J. Lévy Processes. In *Cambridge Tracts in Mathematics*; Cambridge University Press: Cambridge, UK, 1996; Volume 121.

2. Brändle, C.; Colorado, E.; de Pablo, A.; Sánchez, U. A concave-convex elliptic problem involving the fractional Laplacian. *Proc. R. Soc. Edinb.* **2013**, *A143*, 39–71. [CrossRef]

3. Caffarelli, L.; Roquejoffre, J.M.; Sire, Y. Variational problems for free boundaries for the fractional Laplacian. *J. Eur. Math. Soc.* **2010**, *12*, 1151–1179. [CrossRef]

4. Chang, S.Y.A.; González, M. Fractional Laplacian in conformal geometry. *Adv. Math.* **2011**, *226*, 1410–1432. [CrossRef]

5. Di Nezza, E.; Palatucci, G.; Valdinoci, E. Hitchhiker's guide to the fractional sobolev spaces. *Bull. Sci. Math.* **2012**, *136*, 521–573. [CrossRef]

6. Alves, C.O.; Figueiredo, G.M.; Yang, M. Existence of solutions for a nonlinear Choquard equation with potential vanishing at infinity. *Adv. Nonlinear Anal.* **2015**, *5*, 1–15. [CrossRef]

7. Alves, C.O.; Cassani, D.; Tarsi, C.; Yang, M. Existence and concentration of ground state solutions for a critical nonlocal Schrödinger equation in $\mathbb{R}^2$. *J. Differ. Equ.* **2016**, *261*, 1933–1972. [CrossRef]

8. Ding, Y.; Gao, F.; Yang, M. Semiclassical states for Choquard type equations with critical growth: Critical frequency case. *arXiv* **2017**, arXiv:1710.05255.

9. Gao, F.; Yang, M. On the Brezis-Nirenberg type critical problem for nonlinear Choquard equation. *arXiv* **2016**, arXiv:1604.00826.

10. Gao, F.; da Silva, E.D.; Yang, M.; Zhou, J. Existence of solutions for critical Choquard equations via the concentration compactness method. *arXiv* **2017**, arXiv:1712.08264.

11. Lieb, E.H. Existence and uniqueness of the minimizing solution of Choquard's nonlinear equation. *Stud. Appl. Math.* **1977**, *57*, 93–105. [CrossRef]

12. Ma, L.; Zhao, L. Classification of positive solitary solutions of the nonlinear Choquard equation. *Arch. Ration. Mech. Anal.* **2010**, *195*, 455–467. [CrossRef]

13. Moroz, V.; van Schaftingen, J. Existence of groundstates for a class of nonlinear Choquard equations. *Trans. Am. Math. Soc.* **2015**, *367*, 6557–6579. [CrossRef]

14. Shen, Z.; Gao, F.; Yang, M. On critical Choquard equation with potential well. *Discret. Contin. Dyn. Syst. A* **2018**, *38*, 3669–3695. [CrossRef]

15. Moroz, I.M.; Penrose, R.; Tod, P. Spherically-symmetric solutions of the Schrödinger-Newton equations. *Class. Quant. Grav.* **1998**, *15*, 2733–2742. [CrossRef]

16. Pekar, S. *Untersuchung über die Elektronentheorie der Kristalle*; Akademie Verlag: Berlin, Germany, 1954.

17. Frank, R.L.; Lenzmann, E. On ground states for the L^2-critical boson star equation. *arXiv* **2009**, arXiv:0910.2721.

18. D'avenia, P.; Siciliano, G.; Squassina, M. On fractional Choquard equations. *Math. Mod. Meth. Appl. Sci.* **2015**, *25*, 1447–1476. [CrossRef]

19. Shen, Z.; Gao, F.; Yang, M. Ground states for nonlinear fractional Choquard equations with general nonlinearities. *Math. Method. Appl. Sci.* **2016**, *39*, 4082–4098. [CrossRef]

20. Berestycki, H.; Lions, P.-L. Nonlinear scalar field equations, I Existence of a ground state. *Arch. Ration. Mech. Anal.* **1983**, *82*, 313–346. [CrossRef]

21. Alves, C.O.; Souto, M.A.S. Existence of solutions for a class of nonlinear Schrödinger equations with potential vanishing at infinity. *J. Differ. Equ.* **2013**, *254*, 1977–1991. [CrossRef]

22. Barile, S.; Figueiredo, G.M. Existence of least energy positive, negative and nodal solutions for a class of pq-problems with potentials vanishing at infinity. *J. Math. Anal. Appl.* **2015**, *427*, 1205–1233. [CrossRef]
23. Opic, B.; Kufner, A. *Hardy-Type Inequalities*; Pitman Research Notes in Mathematics Series; Longman Scientific and Technical: Harlow, UK, 1990; Volume 219.
24. Szulkin, A.; Weth, T. Ground state solutions for some indefinite variational problems. *J. Funct. Anal.* **2009**, *257*, 3802–3822. [CrossRef]
25. Tang, X. Non-Nehari manifold method for asymptotically periodic Schrödinger equation. *Sci. China Math.* **2015**, *58*, 715–728. [CrossRef]
26. Lieb, E.; Loss, M. *Analysis, Graduate Studies in Mathematics*; AMS: Providence, RI, USA, 2001.
27. Luo, H.; Tang, X. Ground state and multiple solutions for the fractional Schrödinger-Poisson system with critical Sobolev exponent. *Nonlinear Anal. Real World Appl.* **2018**, *42*, 24–52. [CrossRef]
28. Luo, H.; Tang, X.; Gao, Z. Sign-changing solutions for fractional kirchhoff equations in bounded domains. *J. Math. Phys.* **2018**, *59*, 031504. [CrossRef]
29. Willem, M. *Minimax Theorems*; Birkhäuser: Berlin, Germany, 1996.

 *mathematics*

Article

Some Identities on Degenerate Bernstein and Degenerate Euler Polynomials

Taekyun Kim [1,*] and Dae San Kim [2]

[1]   Department of Mathematics, Kwangwoon University, Seoul 139-701, Korea
[2]   Department of Mathematics, Sogang University, Seoul 121-742, Korea; dskim@sogang.ac.kr
*   Correspondence: tkkim@kw.ac.kr

Received: 22 November 2018; Accepted: 3 January 2019; Published: 4 January 2019

Abstract: In recent years, intensive studies on degenerate versions of various special numbers and polynomials have been done by means of generating functions, combinatorial methods, umbral calculus, p-adic analysis and differential equations. The degenerate Bernstein polynomials and operators were recently introduced as degenerate versions of the classical Bernstein polynomials and operators. Herein, we firstly derive some of their basic properties. Secondly, we explore some properties of the degenerate Euler numbers and polynomials and also their relations with the degenerate Bernstein polynomials.

Keywords: degenerate Bernstein polynomials; degenerate Bernstein operators; degenerate Euler polynomials

---

1. Introduction

Let us denote the space of continuous functions on $[0,1]$ by $C[0,1]$, and the space of polynomials of degree $\leq n$ by $\mathbb{P}_n$. The Bernstein operator $\mathbb{B}_n$ of order n, $(n \geq 1)$, associates to each $f \in C[0,1]$ the polynomial $\mathbb{B}_n(f|x) \in \mathbb{P}_n$, and was introduced by Bernstein as (see [1,2]):

$$\mathbb{B}_n(f|x) = \sum_{k=0}^{n} f\left(\frac{k}{n}\right)\binom{n}{k}x^k(1-x)^{n-k}$$
$$= \sum_{k=0}^{n} f\left(\frac{k}{n}\right)B_{k,n}(x), \tag{1}$$

(see [1–14]) where

$$B_{k,n}(x) = \binom{n}{k}x^k(1-x)^{n-k}, \quad (n,k \in \mathbb{Z}_{\geq 0}) \tag{2}$$

are called either Bernstein polynomials of degree n or Bernstein basis polynomials of degree n.

The Bernstein polynomials of degree n can be defined in terms of two such polynomials of degree $n-1$. That is, the k-th Bernstein polynomial of degree n can be written as

$$B_{k,n}(x) = (1-x)B_{k,n-1}(x) + xB_{k-1,n-1}(x), \quad (k,n \in \mathbb{N}). \tag{3}$$

From (2), the first few Bernstein polynomials $B_{k,n}(x)$ are given by

$$B_{0,1}(x) = 1-x, \; B_{1,1}(x) = x, \; B_{0,2}(x) = (1-x)^2, \; B_{1,2}(x) = 2x(1-x),$$
$$B_{2,2}(x) = x^2, \; B_{0,3}(x) = (1-x)^3, \; B_{1,3}(x) = 3x(1-x)^2,$$
$$B_{2,3}(x) = 3x^2(1-x), \; B_{3,3}(x) = x^3, \cdots.$$

Thus, we note that

$$x^k = x(x^{k-1}) = x \sum_{i=k-1}^{n} \frac{\binom{i}{k-1}}{\binom{n}{k-1}} B_{i,n}(x)$$

$$= \sum_{i=k-1}^{n} \frac{\binom{i}{k-1}}{\binom{n}{k-1}} \frac{i+1}{n+1} B_{i+1,n+1}(x)$$

$$= \sum_{i=k-1}^{n} \frac{\binom{i+1}{k}}{\binom{n+1}{k}} B_{i+1,n+1}(x).$$

For $\lambda \in \mathbb{R}$, L. Carlitz introduced the degenerate Euler poynomials given by the generating function (see [15,16])

$$\frac{2}{(1+\lambda t)^{\frac{1}{\lambda}} + 1}(1+\lambda t)^{\frac{x}{\lambda}} = \sum_{n=0}^{\infty} \mathcal{E}_{n,\lambda}(x) \frac{t^n}{n!}, \tag{4}$$

When $x = 0$, $\mathcal{E}_{n,\lambda} = \mathcal{E}_{n,\lambda}(0)$ are called the degenerate Euler numbers. It is easy to show that $\lim_{\lambda \to 0} \mathcal{E}_{n,\lambda}(x) = E_n(x)$, where $E_n(x)$ are the Euler polynomials given by (see [15–18])

$$\frac{2}{e^t + 1} e^{xt} = \sum_{n=0}^{\infty} E_n(x) \frac{t^n}{n!} \tag{5}$$

For $n \geq 0$, we define the λ-product as follows (see [8]):

$$(x)_{0,\lambda} = 1, \quad (x)_{n,\lambda} = x(x - \lambda)(x - 2\lambda) \cdots (x - (n-1)\lambda), \quad (n \geq 1), \tag{6}$$

Observe here that $\lim_{\lambda \to 0}(x)_{n,\lambda} = x^n$, $(n \geq 1)$.
Recently, the degenerate Bernstein polynomials of degree n are introduced as (see [8])

$$B_{k,n}(x|\lambda) = \binom{n}{k}(x)_{k,\lambda}(1-x)_{n-k,\lambda}, \quad (x \in [0,1], \ n,k \geq 0), \tag{7}$$

From (7), it is not difficult to show that the generating function for $B_{k,n}(x|\lambda)$ is given by (see [8])

$$\frac{1}{k!}(x)_{k,\lambda} t^k (1+\lambda t)^{\frac{1-x}{\lambda}} = \sum_{n=k}^{\infty} B_{k,n}(x|\lambda) \frac{t^n}{n!}, \tag{8}$$

By (8), we easily get $\lim_{\lambda \to 0} B_{k,n}(x|\lambda) = B_{k,n}(x)$, $(n,k \geq 0)$.
The Bernstein polynomials are the mathematical basis for Bézier curves which are frequently used in computer graphics and related fields. In this paper, we investigate the degenerate Bernstein polynomials and operators. We study their elementary properties (see also [8]) and then their further properties in association with the degenerate Euler numbers and polynomials.

Finally, we would like to briefly go over some of the recent works related with Bernstein polynomials and operators.

Kim-Kim in Ref. [19] gave identities for degenerate Bernoulli polynomials and Korobov polynomials of the first kind. The authors in Ref. [20] introduced a generalization of the Bernstein polynomials associated with Frobenius–Euler polynomials. The paper [21] deals with some identities of q-Euler numbers and polynomials associated with q-Bernstein polynomials. In Ref. [22], the authors studied a space-time fractional diffusion equation with initial boundary conditions and presented a numerical solution for that. Both normalized Bernstein polynomials with collocation and Galerkin methods are applied to turn the problem into an algebraic system. Kim in Ref. [23] introduced some identities on the q-integral representation of the product of the several q-Bernstein type polynomials. Grouped data are commonly encountered in applications. In Ref. [24], Kim-Kim studied some properties on degenerate Eulerian numbers and polynomials. The authors in Ref. [25] give an overview of several results related to partially degenerate poly-Bernoulli polynomials associated with Hermit polynomials.

2. Degenerate Bernstein Polynomials and Operators

The degenerate Bernstein operator of order n is defined, for $f \in C[0,1]$, as

$$\mathbb{B}_{n,\lambda}(f|x) = \sum_{k=0}^{n} f(\frac{k}{n})\binom{n}{k}(x)_{k,\lambda}(1-x)_{n-k,\lambda} = \sum_{k=0}^{n} f(\frac{k}{n})B_{k,n}(x|\lambda), \tag{9}$$

where $x \in [0,1]$ and $n, k \in \mathbb{Z}_{\geq 0}$.

Theorem 1. *For $n \geq 0$, we have*

$$\mathbb{B}_{n,\lambda}(f|0) = f(0)(1)_{n,\lambda}, \ \mathbb{B}_{n,\lambda}(f|1) = f(1)(1)_{n,\lambda},$$

and

$$\mathbb{B}_{n,\lambda}(1|x) = (1)_{n,\lambda}, \ \mathbb{B}_{n,\lambda}(x|x) = x\sum_{k=0}^{n-1}(-1)^k\lambda^k(n-1)_k(1)_{n-1-k,\lambda}, \ (n \geq 1),$$

where $(x)_k = x(x-1)\cdots(x-k+1), \ (k \geq 1), \ (x)_0 = 1.$

Proof. From (9), we clearly have

$$\mathbb{B}_{n,\lambda}(1|x) = \sum_{k=0}^{n} \binom{n}{k}(x)_{k,\lambda}(1-x)_{n-k,\lambda}. \tag{10}$$

Now, we observe that

$$\sum_{n=0}^{\infty}(1)_{n,\lambda}\frac{t^n}{n!} = (1+\lambda t)^{\frac{1}{\lambda}} = (1+\lambda t)^{\frac{x}{\lambda}}(1+\lambda t)^{\frac{1-x}{\lambda}}$$

$$= \left(\sum_{l=0}^{\infty}(x)_{l,\lambda}\frac{t^l}{l!}\right)\left(\sum_{m=0}^{\infty}(1-x)_{m,\lambda}\frac{t^m}{m!}\right) \tag{11}$$

$$= \sum_{n=0}^{\infty}\left(\sum_{l=0}^{n}\binom{n}{l}(x)_{l,\lambda}(1-x)_{n-l,\lambda}\right)\frac{t^n}{n!}.$$

Comparing the coefficients on both sides of (11), we derive

$$(1)_{n,\lambda} = \sum_{l=0}^{n}\binom{n}{l}(x)_{l,\lambda}(1-x)_{n-l,\lambda}. \tag{12}$$

Combining (10) with (12), we have

$$\mathbb{B}_{n,\lambda}(1|x) = \sum_{k=0}^{n}\binom{n}{k}(x)_{k,\lambda}(1-x)_{n-k,\lambda} = (1)_{n,\lambda}, \ (n \geq 0). \tag{13}$$

Furthermore, we get from (9) that for $f(x) = x$,

$$
\begin{aligned}
\mathbb{B}_{n,\lambda}(x|x) &= \sum_{k=0}^{n} \frac{k}{n} \binom{n}{k} (x)_{k,\lambda} (1-x)_{n-k,\lambda} \\
&= \sum_{k=1}^{n} \binom{n-1}{k-1} (x)_{k,\lambda} (1-x)_{n-k,\lambda} \\
&= \sum_{k=0}^{n-1} \binom{n-1}{k} (x)_{k,\lambda} (1-x)_{n-1-k,\lambda} (x-k\lambda) \\
&= x(1)_{n-1,\lambda} - (n-1)\lambda \sum_{k=0}^{n-1} \binom{n-1}{k} (x)_{k,\lambda} (1-x)_{n-1-k,\lambda} \frac{k}{n-1} \\
&= x(1)_{n-1,\lambda} - (n-1)\lambda \mathbb{B}_{n-1,\lambda}(x|x).
\end{aligned}
\tag{14}
$$

From (14), we can easily deduce the following Equation (15):

$$
\begin{aligned}
\mathbb{B}_{n,\lambda}(x|x) &= x(1)_{n-1,\lambda} - (n-1)\lambda \{ x(1)_{n-2,\lambda} - (n-2)\lambda \mathbb{B}_{n-2,\lambda}(x|x) \} \\
&= x(1)_{n-1,\lambda} - x(n-1)\lambda(1)_{n-2,\lambda} + (-1)^2 (n-1)(n-2)\lambda^2 \mathbb{B}_{n-2,\lambda}(x|x) \\
&= x(1)_{n-1,\lambda} - x(n-1)\lambda(1)_{n-2,\lambda} \\
&\quad + (-1)^2 (n-1)(n-2)\lambda^2 \{ x(1)_{n-3,\lambda} - (n-3)\lambda \mathbb{B}_{n-3,\lambda}(x|x) \} \\
&= x(1)_{n-1,\lambda} - x(n-1)\lambda(1)_{n-2,\lambda} + (-1)^2 (n-1)(n-2)\lambda^2 x(1)_{n-3,\lambda} \\
&\quad + (-1)^3 (n-1)(n-2)(n-3)\lambda^3 \mathbb{B}_{n-3,\lambda}(x|x) \\
&= \cdots \\
&= x \sum_{k=0}^{n-1} (-1)^k \lambda^k (n-1)_k (1)_{n-1-k,\lambda}.
\end{aligned}
\tag{15}
$$

$\square$

Let f, g be continuous functions defined on $[0,1]$. Then, we clearly have

$$
\mathbb{B}_{n,\lambda}(\alpha f + \beta g|x) = \alpha \mathbb{B}_{n,\lambda}(f|x) + \beta \mathbb{B}_{n,\lambda}(g|x), \ (n \geq 0),
\tag{16}
$$

where α, β are constants.

So, the degenerate Bernstein operator is linear. From (7), we note that

$$
\begin{aligned}
&\mathbb{B}_{0,1}(x|\lambda) = 1 - x, \ \mathbb{B}_{1,1}(x|\lambda) = x, \ \mathbb{B}_{0,2}(x|\lambda) = (1-x)^2 - \lambda(1-x), \\
&\mathbb{B}_{1,2}(x|\lambda) = 2x(1-x), \ \mathbb{B}_{2,2}(x|\lambda) = x^2 - \lambda x.
\end{aligned}
$$

It is not hard to see that

$$
\begin{aligned}
\sum_{n=0}^{\infty} (1-x)_{n,\lambda} \frac{t^n}{n!} &= (1+\lambda t)^{\frac{1-x}{\lambda}} = (1+\lambda t)^{\frac{1}{\lambda}} (1+\lambda t)^{-\frac{x}{\lambda}} \\
&= \left(\sum_{l=0}^{\infty} (1)_{l,\lambda} \frac{t^l}{l!} \right) \left(\sum_{m=0}^{\infty} (-x)_{m,\lambda} \frac{t^m}{m!} \right) \\
&= \sum_{n=0}^{\infty} \left(\sum_{l=0}^{n} \binom{n}{l} (1)_{n-l,\lambda} (-1)^l (x)_{l,-\lambda} \right) \frac{t^n}{n!}.
\end{aligned}
$$

This shows that we have

$$
(1-x)_{n,\lambda} = \sum_{l=0}^{n} \binom{n}{l} (1)_{n-l,\lambda} (-1)^l (x)_{l,-\lambda}, \ (n \geq 0).
\tag{17}
$$

Theorem 2. *For $f \in C[0,1]$ and $n \in \mathbb{Z}_{\geq 0}$, we have*

$$\mathbb{B}_{n,\lambda}(f|x) = \sum_{m=0}^{n} \binom{n}{m}(x)_{m,-\lambda} \sum_{k=0}^{m} \binom{m}{k}(-1)^{m-k}(1)_{n-m,\lambda} \frac{(x)_{k,\lambda}}{(x+(m-1)\lambda)_{k,\lambda}} f(\frac{k}{n}).$$

Proof. From (9), it is immediate to see that

$$\mathbb{B}_{n,\lambda}(f|x) = \sum_{k=0}^{n} f(\frac{k}{n}) B_{k,n}(x|\lambda) = \sum_{k=0}^{n} f(\frac{k}{n})\binom{n}{k}(x)_{k,\lambda}(1-x)_{n-k,\lambda}$$

$$= \sum_{k=0}^{n} f(\frac{k}{n})\binom{n}{k}(x)_{k,\lambda} \sum_{j=0}^{n-k} \binom{n-k}{j}(-1)^{j}(1)_{n-k-j,\lambda}(x)_{j,-\lambda}. \tag{18}$$

We need to note the following:

$$\binom{n}{k}\binom{n-k}{j} = \binom{n}{k+j}\binom{k+j}{k}, \quad (n,k \geq 0), \tag{19}$$

and

$$(x)_{j,-\lambda} = \frac{(x)_{k+j,-\lambda}}{(x+(j+k-1)\lambda)_{k,\lambda}}. \tag{20}$$

Let $k+j = m$. Then, by (19), we obviously have

$$\binom{n}{k}\binom{n-k}{j} = \binom{n}{m}\binom{m}{k}. \tag{21}$$

Combining (18) with (19)–(21) gives the following result:

$$\mathbb{B}_{n,\lambda}(f|x) = \sum_{m=0}^{n} \binom{n}{m}(x)_{m,-\lambda} \sum_{k=0}^{m} \binom{m}{k}(-1)^{m-k}(1)_{n-m,\lambda} \frac{(x)_{k,\lambda}}{(x+(m-1)\lambda)_{k,\lambda}} f(\frac{k}{n}).$$

$\square$

Theorem 3. *For $n,k \in \mathbb{Z}_{\geq 0}$ and $x \in [0,1]$, we have*

$$B_{k,n}(x|\lambda) = \sum_{i=k}^{n}(-1)^{i-k}\binom{n}{i}\binom{i}{k}(x)_{i,-\lambda}\frac{(x)_{k,\lambda}}{(x+(i-1)\lambda)_{k,\lambda}}(1)_{n-i,\lambda}.$$

Proof. From (7), (17), and (20), we observe that

$$B_{k,n}(x|\lambda) = \binom{n}{k}(x)_{k,\lambda}(1-x)_{n-k,\lambda}$$

$$= \binom{n}{k}(x)_{k,\lambda} \sum_{i=0}^{n-k} \binom{n-k}{i}(-1)^{i}(x)_{i,-\lambda}(1)_{n-k-i,\lambda}$$

$$= \sum_{i=0}^{n-k}(-1)^{i}\binom{n}{k}\binom{n-k}{i}(x)_{k+i,-\lambda}\frac{(x)_{k,\lambda}}{(x+(k+i-1)\lambda)_{k,\lambda}}(1)_{n-k-i,\lambda} \tag{22}$$

$$= \sum_{i=k}^{n}(-1)^{i-k}\binom{n}{k}\binom{n-k}{i-k}(x)_{i,-\lambda}\frac{(x)_{k,\lambda}}{(x+(i-1)\lambda)_{k,\lambda}}(1)_{n-i,\lambda}$$

$$= \sum_{i=k}^{n}(-1)^{i-k}\binom{n}{i}\binom{i}{k}(x)_{i,-\lambda}\frac{(x)_{k,\lambda}}{(x+(i-1)\lambda)_{k,\lambda}}(1)_{n-i,\lambda}.$$

$\square$

3. Degenerate Euler Polynomials Associated with Degenerate Bernstein Polynomials

Theorem 4. *For $n \geq 0$, the following holds true:*

$$\sum_{l=0}^{n} \binom{n}{l} (1)_{n-l,\lambda} \mathcal{E}_{l,\lambda} + \mathcal{E}_{n,\lambda} = \begin{cases} 2, & \text{if } n = 0, \\ 0, & \text{if } n > 0. \end{cases}$$

Proof. From (4), we remark that

$$\begin{aligned}
2 &= \left(\sum_{l=0}^{\infty} \mathcal{E}_{l,\lambda} \frac{t^l}{l!} \right) \left((1 + \lambda t)^{\frac{1}{\lambda}} + 1 \right) = \left(\sum_{l=0}^{\infty} \mathcal{E}_{l,\lambda} \frac{t^l}{l!} \right) \left(\sum_{m=0}^{\infty} (1)_{m,\lambda} \frac{t^m}{m!} + 1 \right) \\
&= \sum_{n=0}^{\infty} \left(\sum_{l=0}^{n} \binom{n}{l} \mathcal{E}_{l,\lambda} (1)_{n-l,\lambda} \right) \frac{t^n}{n!} + \sum_{n=0}^{\infty} \mathcal{E}_{n,\lambda} \frac{t^n}{n!} \\
&= \sum_{n=0}^{\infty} \left(\sum_{l=0}^{n} \binom{n}{l} (1)_{n-l,\lambda} \mathcal{E}_{l,\lambda} + \mathcal{E}_{n,\lambda} \right) \frac{t^n}{n!}.
\end{aligned} \tag{23}$$

The result follows by comparing the coefficients on both sides of (23). $\square$

From Theorem 4, we note that

$$\mathcal{E}_{0,\lambda} = 1, \quad \mathcal{E}_{n,\lambda} = -\sum_{l=0}^{n} \binom{n}{l} (1)_{n-l,\lambda} \mathcal{E}_{l,\lambda}, \ (n > 0).$$

From these recurrence relations, we note that the first few degenerate Euler numbers are given by

$$\mathcal{E}_{0,\lambda} = 1, \ \mathcal{E}_{1,\lambda} = -\frac{1}{2}, \ \mathcal{E}_{2,\lambda} = \frac{1}{2}\lambda, \ \mathcal{E}_{3,\lambda} = \frac{1}{4} - \lambda^2, \ \mathcal{E}_{4,\lambda} = -\frac{3}{2}\lambda + 3\lambda^3,$$

$$\mathcal{E}_{5,\lambda} = -\frac{1}{2} + \frac{35}{4}\lambda^2 - 12\lambda^4, \ \mathcal{E}_{6,\lambda} = \frac{15}{2}\lambda - \frac{225}{4}\lambda^3 + 60\lambda^5.$$

Theorem 5. *For $n \geq 0$, we have*

$$\mathcal{E}_{n,\lambda}(1 - x) = (-1)^n \mathcal{E}_{n,-\lambda}(x).$$

Especially, we have

$$\mathcal{E}_{n,\lambda}(2) = (-1)^n \mathcal{E}_{n,-\lambda}(-1), \ (n \geq 0).$$

Proof. By (4), we get

$$\begin{aligned}
\sum_{n=0}^{\infty} \mathcal{E}_{n,\lambda}(1 - x) \frac{t^n}{n!} &= \frac{2}{(1 + \lambda t)^{\frac{1}{\lambda}} + 1} (1 + \lambda t)^{\frac{1-x}{\lambda}} = \frac{2}{(1 + \lambda t)^{-\frac{1}{\lambda}} + 1} (1 + \lambda t)^{-\frac{x}{\lambda}} \\
&= \sum_{n=0}^{\infty} \mathcal{E}_{n,-\lambda}(x)(-1)^n \frac{t^n}{n!}.
\end{aligned} \tag{24}$$

Comparing the coefficients on both sides of (24), we have

$$\mathcal{E}_{n,\lambda}(1 - x) = (-1)^n \mathcal{E}_{n,-\lambda}(x), \ (n \geq 0). \tag{25}$$

In particular, if we take $x = -1$, we get

$$\mathcal{E}_{n,\lambda}(2) = (-1)^n \mathcal{E}_{n,-\lambda}(-1), \ (n \geq 0). \tag{26}$$

$\square$

Corollary 1. *For $n \geq 0$, we have*

$$\mathcal{E}_{n,\lambda}(2) = 2(1)_{n,\lambda} + \mathcal{E}_{n,\lambda}.$$

Proof. From (4), we have

$$
\begin{aligned}
\sum_{n=0}^{\infty} \mathcal{E}_{n,\lambda}(2) \frac{t^n}{n!} &= \frac{2}{(1+\lambda t)^{\frac{1}{\lambda}} + 1} (1+\lambda t)^{\frac{2}{\lambda}} \\
&= \frac{2}{(1+\lambda t)^{\frac{1}{\lambda}} + 1} (1+\lambda t)^{\frac{1}{\lambda}} \left((1+\lambda t)^{\frac{1}{\lambda}} + 1 - 1 \right) \\
&= 2(1+\lambda t)^{\frac{1}{\lambda}} - \frac{2}{(1+\lambda t)^{\frac{1}{\lambda}} + 1} (1+\lambda t)^{\frac{1}{\lambda}} \\
&= \sum_{n=0}^{\infty} \left(2(1)_{n,\lambda} - \sum_{l=0}^{n} \binom{n}{l} (1)_{n-l,\lambda} \mathcal{E}_{l,\lambda} \right) \frac{t^n}{n!}.
\end{aligned}
\tag{27}
$$

Thus, by (27), we get

$$\mathcal{E}_{n,\lambda}(2) = 2(1)_{n,\lambda} - \sum_{l=0}^{n} \binom{n}{l} (1)_{n-l,\lambda} \mathcal{E}_{l,\lambda} = 2(1)_{n,\lambda} + \mathcal{E}_{n,\lambda}, \ (n \geq 0).$$

$\square$

Theorem 6. *For $n \geq 0, k \geq 1$, we have*

$$\mathcal{E}_{n,\lambda}(x) = 2 \sum_{i=1}^{k} (-1)^{i-1} (x - i)_{n,\lambda} + (-1)^k \mathcal{E}_{n,\lambda}(x - k).$$

Proof. From (4), we easily see that

$$\mathcal{E}_{n,\lambda}(x) = \sum_{l=0}^{n} \binom{n}{l} \mathcal{E}_{l,\lambda}(x)_{n-l,\lambda}, \ (n \geq 0).$$

Now, we observe that

$$
\frac{2}{(1+\lambda t)^{\frac{1}{\lambda}}+1}(1+\lambda t)^{\frac{x}{\lambda}} = \frac{2}{(1+\lambda t)^{\frac{1}{\lambda}}+1}((1+\lambda t)^{\frac{1}{\lambda}}+1-1)(1+\lambda t)^{\frac{x-1}{\lambda}}
$$

$$
= 2(1+\lambda t)^{\frac{x-1}{\lambda}} - \frac{2}{(1+\lambda t)^{\frac{1}{\lambda}}+1}(1+\lambda t)^{\frac{x-1}{\lambda}}
$$

$$
= 2(1+\lambda t)^{\frac{x-1}{\lambda}} - 2(1+\lambda t)^{\frac{x-2}{\lambda}} + \frac{2}{(1+\lambda t)^{\frac{1}{\lambda}}+1}(1+\lambda t)^{\frac{x-2}{\lambda}}
$$

$$
= 2(1+\lambda t)^{\frac{x-1}{\lambda}} - 2(1+\lambda t)^{\frac{x-2}{\lambda}} + 2(1+\lambda t)^{\frac{x-3}{\lambda}} - \frac{2}{(1+\lambda t)^{\frac{1}{\lambda}}+1}(1+\lambda t)^{\frac{x-3}{\lambda}} \tag{28}
$$

$$
= 2(1+\lambda t)^{\frac{x-1}{\lambda}} - 2(1+\lambda t)^{\frac{x-2}{\lambda}} + 2(1+\lambda t)^{\frac{x-3}{\lambda}} - 2(1+\lambda t)^{\frac{x-4}{\lambda}}
$$

$$
+ \frac{2}{(1+\lambda t)^{\frac{1}{\lambda}}+1}(1+\lambda t)^{\frac{x-4}{\lambda}}.
$$

Continuing the process in (28) gives the following result:

$$
\sum_{n=0}^{\infty} \mathcal{E}_{n,\lambda}(x)\frac{t^n}{n!} = 2\sum_{i=1}^{k}(-1)^{i-1}(1+\lambda t)^{\frac{x-i}{\lambda}} + (-1)^k\frac{2}{(1+\lambda t)^{\frac{1}{\lambda}}+1}(1+\lambda t)^{\frac{x-k}{\lambda}}
$$

$$
= \sum_{n=0}^{\infty}\left(2\sum_{i=1}^{k}(-1)^{i-1}(x-i)_{n,\lambda}\right)\frac{t^n}{n!} + (-1)^k\frac{2}{(1+\lambda t)^{\frac{1}{\lambda}}+1}(1+\lambda t)^{\frac{x-k}{\lambda}}. \tag{29}
$$

The desired result now follows from (4) and (29).

□

Theorem 7. *For $n,k \geq 0$, we have*

$$
B_{k,n+k}(x|\lambda) = \frac{1}{2}(x)_{k,\lambda}\binom{n+k}{k}(\mathcal{E}_{n,\lambda}(2-x) + \mathcal{E}_{n,\lambda}(1-x)).
$$

Proof. In view of (8), we have

$$
(x)_{k,\lambda}(1+\lambda t)^{\frac{1-x}{\lambda}} = \frac{k!}{t^k}\sum_{n=k}^{\infty}B_{k,n}(x|\lambda)\frac{t^n}{n!} = \sum_{n=0}^{\infty}B_{k,n+k}(x|\lambda)\frac{1}{\binom{n+k}{n}}\frac{t^n}{n!}. \tag{30}
$$

On the other hand, (30) is also given by

$$
(x)_{k,\lambda}(1+\lambda t)^{\frac{1-x}{\lambda}} = \sum_{n=0}^{\infty}(x)_{k,\lambda}(1-x)_{n,\lambda}\frac{t^n}{n!}. \tag{31}
$$

From (30) and (31), we have

$$
(x)_{k,\lambda}(1-x)_{n,\lambda} = \frac{1}{\binom{n+k}{n}}B_{k,n+k}(x|\lambda), \quad (n,k \geq 0). \tag{32}
$$

Now, we observe that

$$
(x)_{k,\lambda}(1+\lambda t)^{\frac{1-x}{\lambda}} = \frac{(x)_{k,\lambda}}{2}\frac{2}{(1+\lambda t)^{\frac{1}{\lambda}}+1}(1+\lambda t)^{\frac{1-x}{\lambda}}((1+\lambda t)^{\frac{1}{\lambda}}+1)
$$

$$
= \frac{(x)_{k,\lambda}}{2}\left(\frac{2}{(1+\lambda t)^{\frac{1}{\lambda}}+1}(1+\lambda t)^{\frac{2-x}{\lambda}}+\frac{2}{(1+\lambda t)^{\frac{1}{\lambda}}+1}(1+\lambda t)^{\frac{1-x}{\lambda}}\right) \tag{33}
$$

$$
= \frac{(x)_{k,\lambda}}{2}\left(\sum_{n=0}^{\infty}(\mathcal{E}_{n,\lambda}(2-x)+\mathcal{E}_{n,\lambda}(1-x))\frac{t^n}{n!}\right).
$$

By (31) and (33), we get

$$
(x)_{k,\lambda}(1-x)_{n,\lambda} = \frac{(x)_{k,\lambda}}{2}(\mathcal{E}_{n,\lambda}(2-x)+\mathcal{E}_{n,\lambda}(1-x)),\ (n\geq 0). \tag{34}
$$

Therefore, from (32) and (34), we have the result. $\square$

4. Conclusions

In 1912, Bernstein first used Bernstein polynomials to give a constructive proof for the Stone–Weierstrass approximation theorem. The convergence of the Bernstein approximation of a function f to f is of order $1/n$, even for smooth functions, and hence the related approximation process is not used for computational purposes. However, by combining Bernstein approximations and the use of ad hoc extrapolation algorithms, fast techniques were designed (see the recent review [14] and paper [13]). Furthermore, about half a century later, they were used to design automobile bodies at Renault by Pierre Bézier. The Bernstein polynomials are the mathematical basis for Bézier curves, which are frequently used in computer graphics and related fields such as animation, modeling, CAD, and CAGD.

The study of degenerate versions of special numbers and polynomials began with the papers by Carlitz in Refs. [15,16]. Kim and his colleagues have been studying various degenerate numbers and polynomials by means of generating functions, combinatorial methods, umbral calculus, p-adic analysis, and differential equations. This line of study led even to the introduction to degenerate gamma functions and degenerate Laplace transforms (see [26]). These already demonstrate that studying degenerate versions of known special numbers and polynomials can be very promising and rewarding. Furthermore, we can hope that many applications will be found not only in mathematics but also in sciences and engineering. As we mentioned in the above, it was not until about fifty years later that Bernstein polynomials found their applications in real-world problems.

With this hope in mind, here we investigated the degenerate Bernstein polynomials and operators which were recently introduced as degenerate versions of the classical Bernstein polynomials and operators. We derived some of their basic properties. In addition, we studied some further properties of the degenerate Bernstein polynomials related to the degenerate Euler numbers and polynomials.

Author Contributions: T.K. and D.S.K. conceived the framework and structured the whole paper; T.K. wrote the paper; D.S.K. checked the results of the paper; D.S.K. and T. K. completed the revision of the article.

Funding: This research received no external funding.

Acknowledgments: The first author's work in this paper was conducted during the sabbatical year of Kwangwoon University in 2018. The authors would like to express their sincere gratitude to the referees for their valuable comments, which have improved the presentation of this paper.

Conflicts of Interest: The authors declare no conflict of interest.

References

1. Kurt, V. Some relation between the Bernstein polynomials and second kind Bernoulli polynomials. *Adv. Stud. Contemp. Math. (Kyungshang)* **2013**, *23*, 43–48.

2.   Kim, T. Identities on the weighted q–Euler numbers and q-Bernstein polynomials. *Adv. Stud. Contemp. Math. (Kyungshang)* **2012**, *22* , 7–12. [CrossRef]

3.   Kim, T. A note on q–Bernstein polynomials. *Russ. J. Math. Phys.* **2011**, *18* , 73–82. [CrossRef]

4.   Kim, T. A study on the q–Euler numbers and the fermionic q–integral of the product of several type q–Bernstein polynomials on $\mathbb{Z}_p$. *Adv. Stud. Contemp. Math. (Kyungshang)* **2013**, *23*, 5–11.

5.   Kim, T.; Bayad, A.; Kim, Y.H. A study on the p–adic q–integral representation on $\mathbb{Z}_p$ associated with the weighted q–Bernstein and q–Bernoulli polynomials. *J. Inequal. Appl.* **2011**, *2011*, 513821. [CrossRef]

6.   Kim, T.; Choi, J.; Kim, Y.-H. On the k–dimensional generalization of q–Bernstein polynomials. *Proc. Jangjeon Math. Soc.* **2011**, *14*, 199–207.

7.   Kim, T.; Choi, J.; Kim, Y.H. Some identities on the q–Bernstein polynomials, q–Stirling numbers and q-Bernoulli numbers. *Adv. Stud. Contemp. Math. (Kyungshang)* **2010**, *20*, 335–341.

8.   Kim, T.; Kim, D.S. Degenerate Bernstein Polynomials. *RACSAM* **2018**. [CrossRef]

9.   Kim, T.; Lee, B.; Choi, J.; Kim, Y.H.; Rim, S.H. On the q–Euler numbers and weighted q–Bernstein polynomials. *Adv. Stud. Contemp. Math. (Kyungshang)* **2011**, *21*, 13–18.

10.   Rim, S.-H.; Joung, J.; Jin, J.-H.; Lee, S.-J. A note on the weighted Carlitz's type q=Euler numbers and q-Bernstein polynomials. *Proc. Jangjeon Math. Soc.* **2012**, *15*, 195–201.

11.   Simsek, Y. Identities and relations related to combinatorial numbers and polynomials. *Proc. Jangjeon Math. Soc.* **2017**, *20*, 127–135.

12.   Siddiqui, M.A.; Agarwal, R.R.; Gupta, N. On a class of modified new Bernstein operators. *Adv. Stud. Contemp. Math. (Kyungshang)* **2014**, *24*, 97–107.

13.   Costabile, F.; Gualtieri, M.; Serra, S. Asymptotic expansion and extrapolation for Bernstein polynomials with applications. *BIT Numer. Math.* **1996**, *36*, 676–687. [CrossRef]

14.   Khosravian, H.; Dehghan, M.; Eslahchi, M.R. A new approach to improve the order of approximation of the Bernstein operators: Theory and applications. *Numer. Algorithms* **2018**, *77*, 111–150. [CrossRef]

15.   Carlitz, L. Degenerate Stirling, Bernoulli and Eulerian numbers. *Utilitas Math.* **1979**, *15*, 51–88.

16.   Carlitz, L. A degenerate Staudt-Clausen theorem. *Arch. Math.* **1956**, *7*, 28–33. [CrossRef]

17.   Bayad, A.; Kim, T.; Lee, B.; Rim, S.-H. Some identities on Bernstein polynomials associated with q-Euler polynomials. *Abstr. Appl. Anal.* **2011**, *2011*, 294715. [CrossRef]

18.   Bernstein, S. Demonstration du theoreme de Weierstrass basee sur le calcul des probabilites. *Commun. Soc. Math. Kharkow* **1912**, *13*, 1–2.

19.   Kim, T.; Kim, D.S. Identities for degenerate Bernoulli polynomials and Korobov polynomials of the first kind. *Sci. China Math.* **2018**. [CrossRef]

20.   Araci, S.; Acikgoz, M. A note on the Frobenius-Euler numbers and polynomials associated with Bernstein polynomials. *Adv. Stud. Contemp. Math.* **2012**, *22*, 399–406.

21.   Ryoo, C.S. Some identities of the twisted q-Euler numbers and polynomials associated with q-Bernstein polynomials. *Proc. Jangjeon Math. Soc.* **2011**, *14*, 239–248.

22.   Bayad, A.; Kim, T. Identities involving values of Bernstein, q–Bernoulli, and q–Euler polynomials. *Russ. J. Math. Phys.* **2011**, *18*, 133–143. [CrossRef]

23.   Kim, T. Some identities on the q-integral representation of the product of several q-Bernstein-type polynomials. *Abstr. Appl. Anal.* **2011**, *2011*, 634675. [CrossRef]

24.   Kim, D.S.; Kim, T. A note on degenerate Eulerian numbers and polynomials. *Adv. Stud. Contemp. Math. (Kyungshang)* **2017**, *27*, 431–440.

25.   Khan, W.A.; Ahmad, M. Partially degenerate poly-Bernoulli polynomials associated with Hermite polynomials. *Adv. Stud. Contemp. Math. (Kyungshang)* **2018**, *28*, 487–496.

26.   Kim, T.; Kim, D.S. Degenerate Laplace transform and degenerate gamma function. *Russ. J. Math. Phys.* **2017**, *24*, 241–248. [CrossRef]

 *mathematics* 

Article

Some Identities Involving Hermite Kampé de Fériet Polynomials Arising from Differential Equations and Location of Their Zeros

Cheon Seoung Ryoo

Department of Mathematics, Hannam University, Daejeon 306-791, Korea; ryoocs@hnu.kr

Received: 27 November 2018 ; Accepted: 24 December 2018; Published: 26 December 2018 

Abstract: In this paper, we study differential equations arising from the generating functions of Hermit Kampé de Fériet polynomials. Use this differential equation to give explicit identities for Hermite Kampé de Fériet polynomials. Finally, use the computer to view the location of the zeros of Hermite Kampé de Fériet polynomials.

Keywords: differential equations, heat equation; Hermite Kampé de Fériet polynomials; Hermite polynomials; generating functions; complex zeros

2000 Mathematics Subject Classification: 05A19; 11B83; 34A30; 65L99

---

1. Introduction

Numerous studies have been conducted on Bernoulli polynomials, Euler polynomials, tangent polynomials, Hermite polynomials and Laguerre polynomials (see [1–13]). The special polynomials of the two variables provided a new way to analyze solutions of various kinds of partial differential equations that are often encountered in physical problems. Most of the special function of mathematical physics and their generalization have been proposed as physical problems. For example, we recall that the two variables Hermite Kampé de Fériet polynomials $H_n(x, y)$ defined by the generating function (see [2])

$$\sum_{n=0}^{\infty} H_n(x, y) \frac{t^n}{n!} = e^{xt + yt^2} = F(t, x, y) \tag{1}$$

are the solution of heat equation

$$\frac{\partial}{\partial y} H_n(x, y) = \frac{\partial^2}{\partial x^2} H_n(x, y), \quad H_n(x, 0) = x^n.$$

We note that $H_n(2x, -1) = H_n(x)$, where $H_n(x)$ are the classical Hermite polynomials (see [1]). The differential equation and relation are given by

$$\left(2y \frac{\partial^2}{\partial x^2} + x \frac{\partial}{\partial x} - n \right) H_n(x, y) = 0 \text{ and } \frac{\partial}{\partial y} H_n(x, y) = \frac{\partial^2}{\partial x^2} H_n(x, y),$$

respectively.

By (1) and Cauchy product, we get

$$\sum_{n=0}^{\infty} H_n(x_1 + x_2, y) \frac{t^n}{n!} = e^{(x_1+x_2)t+yt^2}$$

$$= \sum_{n=0}^{\infty} H_n(x_1, y) \frac{t^n}{n!} \sum_{n=0}^{\infty} x_2^n \frac{t^n}{n!} \qquad (2)$$

$$= \sum_{n=0}^{\infty} \left(\sum_{l=0}^{n} \binom{n}{l} H_l(x_1, y) x_2^{n-l} \right) \frac{t^n}{n!}.$$

By comparing the coefficients on both sides of (2), we have the following theorem:

Theorem 1. *For any positive integer n, we have*

$$H_n(x_1 + x_2, y) = \sum_{l=0}^{n} \binom{n}{l} H_l(x_1, y) x_2^{n-l}.$$

The following elementary properties of the two variables Hermite Kampé de Fériet polynomials $H_n(x, y)$ are readily derived from (1).

Theorem 2. *For any positive integer n, we have*

$$(1) \quad H_n(x, y_1 + y_2) = n! \sum_{l=0}^{\left[\frac{n}{2}\right]} \frac{H_{n-2l}(x, y_1) y_2^l}{l!(n - 2l)!},$$

$$(2) \quad H_n(x, y) = \sum_{l=0}^{n} \binom{n}{l} H_l(x) H_{n-l}(-x, y + 1),$$

$$(3) \quad H_n(x_1 + x_2, y_1 + y_2) = \sum_{l=0}^{n} \binom{n}{l} H_l(x_1, y_1) H_{n-l}(x_2, y_2).$$

Recently, many mathematicians have studied differential equations that occur in the generating functions of special polynomials (see [8,9,14–16]). The paper is organized as follows. We derive the differential equations generated from the generating function of Hermite Kampé de Fériet polynomials:

$$\left(\frac{\partial}{\partial t} \right)^N F(t, x, y) - a_0(N, x, y) F(t, x, y) - \cdots - a_N(N, x, y) t^N F(t, x, y) = 0.$$

By obtaining the coefficients of this differential equation, we obtain explicit identities for the Hermite Kampé de Fériet polynomials in Section 2. In Section 3, we investigate the zeros of the Hermite Kampé de Fériet polynomials using numerical methods. Finally, we observe the scattering phenomenon of the zeros of Hermite Kampé de Fériet polynomials.

2. Differential Equations Associated with Hermite Kampé de Fériet Polynomials

In order to obtain explicit identities for special polynomials, differential equations arising from the generating functions of special polynomials are studied by many authors (see [8,9,14–16]). In this section, we introduce differential equations arising from the generating functions of Hermite Kampé de Fériet polynomials and use these differential equations to obtain the explicit identities for the Hermite Kampé de Fériet polynomials.

Let

$$F = F(t, x, y) = e^{xt+yt^2} = \sum_{n=0}^{\infty} H_n(x, y) \frac{t^n}{n!}, \quad x, y, t \in \mathbb{C}. \qquad (3)$$

Then, by (3), we have

$$F^{(1)} = \frac{\partial}{\partial t}F(t,x,y) = \frac{\partial}{\partial t}\left(e^{xt+yt^2}\right) = e^{xt+yt^2}(x+2yt)$$
$$= (x+2yt)F(t,x,y),$$

$$F^{(2)} = \frac{\partial}{\partial t}F^{(1)}(t,x,y) = 2yF(t,x,y) + (x+2yt)F^{(1)}(t,x,y)$$
$$= (2y + x^2 + (4xy)t + 4y^2t^2)F(t,x,y),$$

and

$$F^{(3)} = \frac{\partial}{\partial t}F^{(2)}(t,x,y)$$
$$= (4xy + 8y^2t)F(t,x,y) + (2y + x^2 + (4xy)t + 4y^2t^2)F^{(1)}(t,x,y)$$
$$= (6xy + x^3)F^{(2)}(t,x,y)$$
$$+ (8y^2 + 4x^2y + 4y^2 + 2x^2y)tF(t,x,y)$$
$$+ (4xy^2 + 8xy^2)t^2F(t,x,y).$$

If we continue this process, we can guess as follows:

$$F^{(N)} = \left(\frac{\partial}{\partial t}\right)^N F(t,x,y) = \sum_{i=0}^{N} a_i(N,x,y)t^i F(t,x,y), \quad (N = 0,1,2,\ldots). \tag{4}$$

Differentiating (4) with respect to t, we have

$$F^{(N+1)} = \frac{\partial F^{(N)}}{\partial t}$$
$$= \sum_{i=0}^{N} a_i(N,x,y)it^{i-1}F(t,x,y) + \sum_{i=0}^{N} a_i(N,x,y)t^i F^{(1)}(t,x,y)$$
$$= \sum_{i=0}^{N} a_i(N,x,y)it^{i-1}F(t,x,y) + \sum_{i=0}^{N} a_i(N,x,y)t^i(x+2yt)F(t,x,y)$$
$$= \sum_{i=0}^{N} ia_i(N,x,y)t^{i-1}F(t,x,y) + \sum_{i=0}^{N} xa_i(N,x,y)t^i F(t,x,y) \tag{5}$$
$$+ \sum_{i=0}^{N} 2ya_i(N,x,y)t^{i+1}F(t,x,y)$$
$$= \sum_{i=0}^{N-1} (i+1)a_{i+1}(N,x,y)t^i F(t,x,y) + \sum_{i=0}^{N} xa_i(N,x,y)t^i F(t,x,y)$$
$$+ \sum_{i=1}^{N+1} 2ya_{i-1}(N,x,y)t^i F(t,x,y).$$

Now, replacing N by $N+1$ in (4), we find

$$F^{(N+1)} = \sum_{i=0}^{N+1} a_i(N+1,x,y)t^i F(t,x,y). \tag{6}$$

Comparing the coefficients on both sides of (5) and (6), we obtain

$$a_0(N+1,x,y) = a_1(N,x,y) + xa_0(N,x,y),$$
$$a_N(N+1,x,y) = xa_N(N,x,y) + 2ya_{N-1}(N,x,y), \tag{7}$$
$$a_{N+1}(N+1,x,y) = 2ya_N(N,x,y),$$

and

$$a_i(N+1,x,y) = (i+1)a_{i+1}(N,x,y) + xa_i(N,x,y) + 2ya_{i-1}(N,x,y), (1 \leq i \leq N-1). \tag{8}$$

In addition, by (4), we have

$$F(t,x,y) = F^{(0)}(t,x,y) = a_0(0,x,y)F(t,x,y), \tag{9}$$

which gives

$$a_0(0,x,y) = 1. \tag{10}$$

It is not difficult to show that

$$\begin{aligned}
xF(t,x,y) &+ 2ytF(t,x,y) \\
&= F^{(1)}(t,x,y) \\
&= \sum_{i=0}^{1} a_i(1,x,y)F(t,x,y) \\
&= a_0(1,x,y)F(t,x,y) + a_1(1,x,y)tF(t,x,y).
\end{aligned} \tag{11}$$

Thus, by (11), we also find

$$a_0(1,x,y) = x, \quad a_1(1,x,y) = 2y. \tag{12}$$

From (7), we note that

$$\begin{aligned}
a_0(N+1,x,y) &= a_1(N,x,y) + xa_0(N,x,y), \\
a_0(N,x,y) &= a_1(N-1,x,y) + xa_0(N-1,x,y), \ldots \\
a_0(N+1,x,y) &= \sum_{i=0}^{N} x^i a_1(N-i,x,y) + x^{N+1},
\end{aligned} \tag{13}$$

$$\begin{aligned}
a_N(N+1,x,y) &= xa_N(N,x,y) + 2ya_{N-1}(N,x,y), \\
a_{N-1}(N,x,y) &= xa_{N-1}(N-1,x,y) + 2ya_{N-2}(N-1,x,y), \ldots \\
a_N(N+1,x,y) &= (N+1)x(2y)^N,
\end{aligned} \tag{14}$$

and

$$\begin{aligned}
a_{N+1}(N+1,x,y) &= 2ya_N(N,x,y), \\
a_N(N,x,y) &= 2ya_{N-1}(N-1,x,y), \ldots \\
a_{N+1}(N+1,x,y) &= (2y)^{N+1}.
\end{aligned} \tag{15}$$

For $i = 1$ in (8), we have

$$a_1(N+1,x,y) = 2\sum_{k=0}^{N} x^k a_2(N-k,x,y) + (2y)\sum_{k=0}^{N} x^k a_0(N-k,x,y). \tag{16}$$

Continuing this process, we can deduce that, for $1 \leq i \leq N-1$,

$$a_i(N+1,x,y) = (i+1)\sum_{k=0}^{N} x^k a_{i+1}(N-k,x,y) + (2y)\sum_{k=0}^{N} x^k a_{i-1}(N-k,x,y). \tag{17}$$

Note that here the matrix $a_i(j, x, y)_{0 \le i,j \le N+1}$ is given by

$$
\begin{pmatrix}
1 & x & 2y + x^2 & 6xy + x^3 & \cdots & & \cdot \\
0 & 2y & 2x(2y) & \cdot & \cdots & & \cdot \\
0 & 0 & (2y)^2 & 3x(2y)^2 & \cdots & & \cdot \\
0 & 0 & 0 & (2y)^3 & \ddots & & \cdot \\
\vdots & \vdots & \vdots & \vdots & & \ddots & (N+1)x(2y)^N \\
0 & 0 & 0 & 0 & & \cdots & (2y)^{N+1}
\end{pmatrix}
$$

Therefore, we obtain the following theorem.

Theorem 3. *For $N = 0, 1, 2, \ldots$, the differential equation*

$$
F^{(N)} = \left(\frac{\partial}{\partial t} \right)^N F(t, x, y) = \left(\sum_{i=0}^{N} a_i(N, x, y) t^i \right) F(t, x, y)
$$

has a solution

$$
F = F(t, x, y) = e^{xt + yt^2},
$$

where

$$
a_0(N, x, y) = \sum_{k=0}^{N-1} x^i a_1(N - 1 - k, x, y) + x^N,
$$

$$
a_{N-1}(N, x, y) = Nx(2y)^{N-1},
$$
$$
a_N(N, x, y) = (2y)^N,
$$

$$
a_i(N + 1, x, y) = (i + 1) \sum_{k=0}^{N} x^k a_{i+1}(N - k, x, y) + (2y) \sum_{k=0}^{N} x^k a_{i-1}(N - k, x, y),
$$

$$
(1 \le i \le N - 2).
$$

Making N-times derivative for (3) with respect to t, we have

$$
\left(\frac{\partial}{\partial t} \right)^N F(t, x, y) = \left(\frac{\partial}{\partial t} \right)^N e^{xt + yt^2} = \sum_{m=0}^{\infty} H_{m+N}(x, y) \frac{t^m}{m!}. \tag{18}
$$

By Cauchy product and multiplying the exponential series $e^{xt} = \sum_{m=0}^{\infty} x^m \dfrac{t^m}{m!}$ in both sides of (18), we get

$$
e^{-nt} \left(\frac{\partial}{\partial t} \right)^N F(t, x, y) = \left(\sum_{m=0}^{\infty} (-n)^m \frac{t^m}{m!} \right) \left(\sum_{m=0}^{\infty} H_{m+N}(x, y) \frac{t^m}{m!} \right)
$$
$$
= \sum_{m=0}^{\infty} \left(\sum_{k=0}^{m} \binom{m}{k} (-n)^{m-k} H_{N+k}(x, y) \right) \frac{t^m}{m!}. \tag{19}
$$

For non-negative integer m, assume that $\{a(m)\}, \{b(m)\}, \{c(m)\}, \{\bar{c}(m)\}$ are four sequences given by

$$
\sum_{m=0}^{\infty} a(m) \frac{t^n}{m!}, \quad \sum_{m=0}^{\infty} b(m) \frac{t^m}{m!}, \quad \sum_{m=0}^{\infty} c(m) \frac{t^m}{m!}, \quad \sum_{m=0}^{\infty} \bar{c}(m) \frac{t^m}{m!}.
$$

If $\sum_{m=0}^{\infty} c(m)\dfrac{t^m}{m!} \times \sum_{m=0}^{\infty} \bar{c}(m)\dfrac{t^m}{m!} = 1$, we have the following inverse relation:

$$a(m) = \sum_{k=0}^{m} \binom{m}{k} c(k)b(m-k) \iff b(m) = \sum_{k=0}^{m} \binom{m}{k} \bar{c}(k)a(m-k). \tag{20}$$

By (20) and the Leibniz rule, we have

$$\begin{aligned}
e^{-nt}\left(\frac{\partial}{\partial t}\right)^N F(t,x,y) &= \sum_{k=0}^{N} \binom{N}{k} n^{N-k}\left(\frac{\partial}{\partial t}\right)^k \left(e^{-nt}F(t,x,y)\right) \\
&= \sum_{m=0}^{\infty}\left(\sum_{k=0}^{N}\binom{N}{k}n^{N-k}H_{m+k}(x-n,y)\right)\frac{t^m}{m!}.
\end{aligned} \tag{21}$$

Hence, by (19) and (21), and comparing the coefficients of $\dfrac{t^m}{m!}$ gives the following theorem.

Theorem 4. *Let m, n, N be nonnegative integers. Then,*

$$\sum_{k=0}^{m} \binom{m}{k}(-n)^{m-k}H_{N+k}(x,y) = \sum_{k=0}^{N} \binom{N}{k}n^{N-k}H_{m+k}(x-n,y). \tag{22}$$

If we take $m = 0$ in (22), then we have the following:

Corollary 1. *For $N = 0, 1, 2, \ldots$, we have*

$$H_N(x,y) = \sum_{k=0}^{N} \binom{N}{k}n^{N-k}H_k(x-n,y).$$

For $N = 0, 1, 2, \ldots$, the differential equation

$$F^{(N)} = \left(\frac{\partial}{\partial t}\right)^N F(t,x,y) = \left(\sum_{i=0}^{N} a_i(N,x,y)t^i\right) F(t,x,y)$$

has a solution

$$F = F(t,x,y) = e^{xt+yt^2}.$$

Here is a plot of the surface for this solution.

In Figure 1 (left), we choose $-3 \le x \le 3, -1 \le t \le 1$, and $y = 3$. In Figure 1 (right), we choose $-3 \le x \le 3, -1 \le t \le 1$, and $y = -3$.

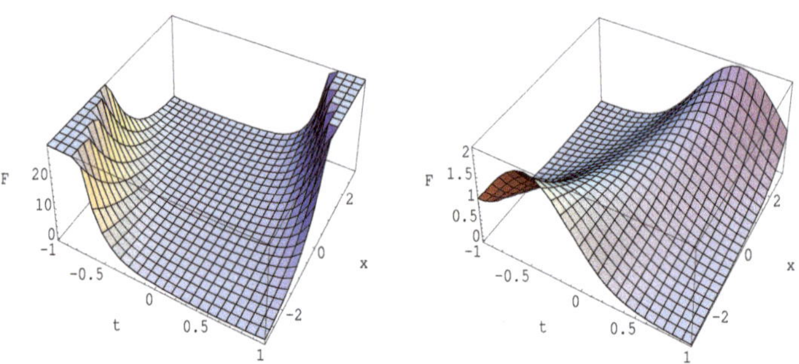

Figure 1. The surface for the solution $F(t,x,y)$.

3. Zeros of the Hermite Kampé de Fériet Polynomials

By using software programs, many mathematicians can explore concepts more easily than in the past. These experiments allow mathematicians to quickly create and visualize new ideas, review properties of figures, create many problems, and find and guess patterns. This numerical survey is particularly interesting because it helps many mathematicians understand basic concepts and solve problems. In this section, we examine the distribution and pattern of zeros of Hermite Kampé de Fériet polynomials $H_n(x,y)$ according to the change of degree n. Based on these results, we present a problem that needs to be approached theoretically.

By using a computer, the Hermite Kampé de Fériet polynomials $H_n(x,y)$ can be determined explicitly. First, a few examples of them are as follows:

$$H_0(x,y) = 1,$$
$$H_1(x,y) = x,$$
$$H_2(x,y) = x^2 + 2y,$$
$$H_3(x,y) = x^3 + 6xy,$$
$$H_4(x,y) = x^4 + 12x^2y + 12y^2,$$
$$H_5(x,y) = x^5 + 20x^3y + 60xy^2,$$
$$H_6(x,y) = x^6 + 30x^4y + 180x^2y^2 + 120y^3,$$
$$H_7(x,y) = x^7 + 42x^5y + 420x^3y^2 + 840xy^3,$$
$$H_8(x,y) = x^8 + 56x^6y + 840x^4y^2 + 3360x^2y^3 + 1680y^4,$$
$$H_9(x,y) = x^9 + 72x^7y + 1512x^5y^2 + 10,080x^3y^3 + 15,120xy^4,$$
$$H_{10}(x,y) = x^{10} + 90x^8y + 2520x^6y^2 + 25,200x^4y^3 + 75,600x^2y^4 + 30,240y^5.$$

Using a computer, we investigate the distribution of zeros of the Hermite Kampé de Fériet polynomials $H_n(x,y)$.

Plots the zeros of the polynomial $H_n(x,y)$ for $n = 20, y = 2, -2, 2 + i, -2 + i$ and $x \in \mathbb{C}$ are as follows (Figure 1). In Figure 2 (top-left), we choose $n = 20$ and $y = 2$. In Figure 2 (top-right), we choose $n = 20$ and $y = -2$. In Figure 2 (bottom-left), we choose $n = 20$ and $y = 2 + i$. In Figure 2 (bottom-right), we choose $n = 20$ and $y = -2 - i$.

Stacks of zeros of the Hermite Kampé de Fériet polynomials $H_n(x,y)$ for $1 \leq n \leq 20$ from a 3D structure are presented (Figure 3). In Figure 3 (top-left), we choose $y = 2$. In Figure 3 (top-right), we choose $y = -2$. In Figure 3 (bottom-left), we choose $y = 2 + i$. In Figure 3 (bottom-right), we choose $y = -2 - i$. Our numerical results for approximate solutions of real zeros of the Hermite Kampé de Fériet polynomials $H_n(x,y)$ are displayed (Tables 1–3).

The plot of real zeros of the Hermite Kampé de Fériet polynomials $H_n(x,y)$ for $1 \leq n \leq 20$ structure are presented (Figure 4). It is expected that $H_n(x,y), x \in \mathbb{C}, y > 0$, has $Im(x) = 0$ reflection symmetry analytic complex functions (see Figures 2 and 3). We also expect that $H_n(x,y), x \in \mathbb{C}, y < 0$, has $Re(x) = 0$ reflection symmetry analytic complex functions (see Figures 2–4). We observe a remarkable regular structure of the complex roots of the Hermite Kampé de Fériet polynomials $H_n(x,y)$ for $y < 0$. We also hope to verify a remarkable regular structure of the complex roots of the Hermite Kampé de Fériet polynomials $H_n(x,y)$ for $y < 0$ (Table 1). Next, we calculated an approximate solution that satisfies $H_n(x,y) = 0, x \in \mathbb{C}$. The results are shown in Table 3.

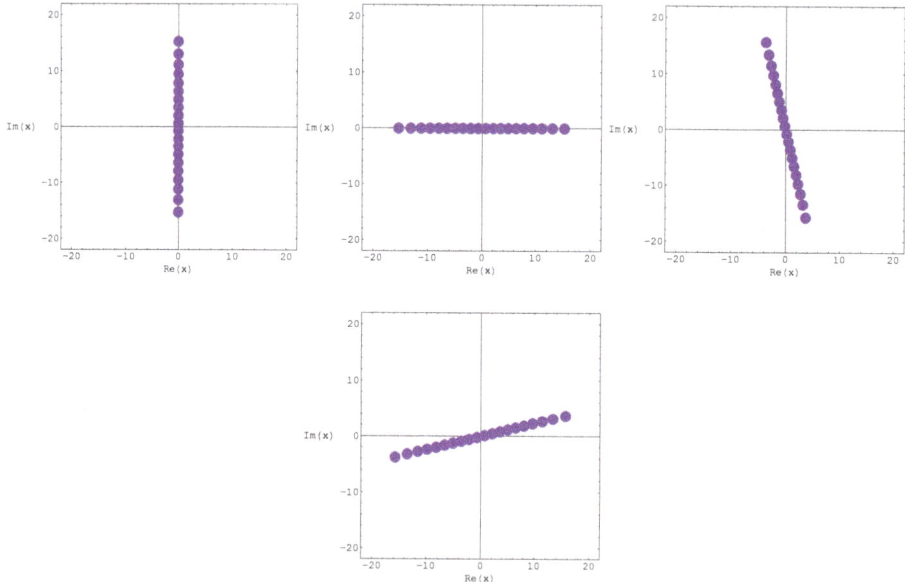

Figure 2. Zeros of $H_n(x, y)$.

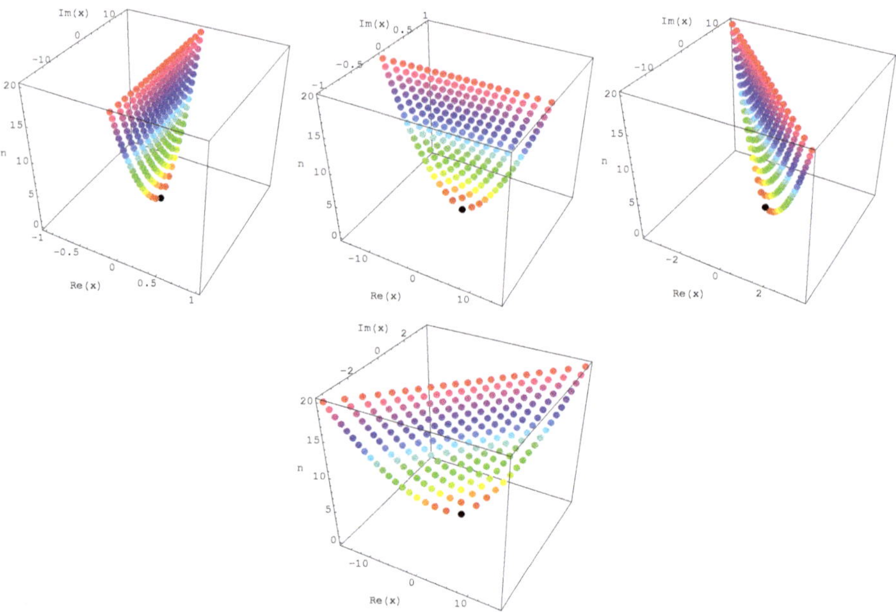

Figure 3. Stacks of zeros of $H_n(x, y)$, $1 \leq n \leq 20$.

Table 1. Numbers of real and complex zeros of $H_n(x, -2)$.

Degree n	Real Zeros	Complex Zeros
1	1	0
2	2	0
3	3	0
4	4	0
5	5	0
6	6	0
7	7	0
8	8	0
9	9	0
10	10	0
11	11	0
12	12	0
13	13	0
14	14	0
⋮	⋮	⋮
29	29	0
30	30	0

Table 2. Numbers of real and complex zeros of $H_n(x, 2)$.

Degree n	Real Zeros	Complex Zeros
1	0	1
2	0	2
3	0	3
4	0	4
5	0	5
6	0	6
7	0	7
8	0	8
9	0	9
10	0	10
11	0	11
12	0	12
13	0	13
14	0	14
⋮	⋮	⋮
29	0	29
30	0	30

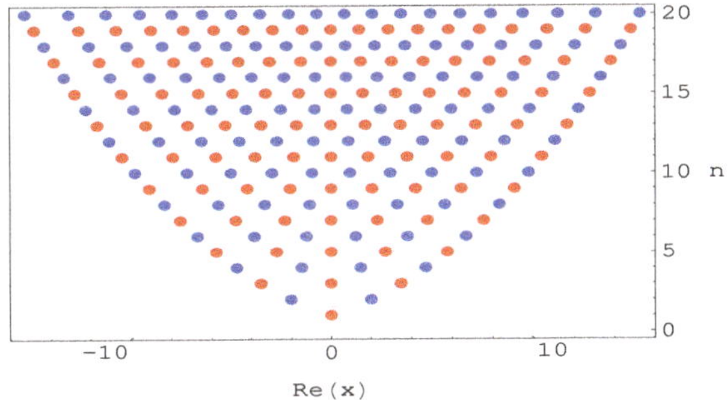

Figure 4. Real zeros of $H_n(x, -2)$ for $1 \leq n \leq 20$.

Table 3. Approximate solutions of $H_n(x, -2) = 0, x \in \mathbb{R}$.

Degree n	x
1	0
2	−2.0000, 2.0000
3	−3.4641, 0, 3.4641
4	−4.669, −1.4839, 1.4839, 4.669
5	−5.714, −2.711, 0, 2.711, 5.714
6	−6.65, −3.778, −1.233, 1.233, 3.778, 6.65
7	−7.50, −4.73, −2.309, 0, 2.309, 4.73, 7.50
8	−8.3, −5.6, −3.27, −1.078, 1.078, 3.27, 5.6, 8.3

4. Conclusions and Future Developments

This study obtained the explicit identities for Hermite Kampé de Fériet polynomials $H_n(x, y)$. The location and symmetry of the roots of the Hermite Kampé de Fériet polynomials were investigated. We examined the symmetry of the zeros of the Hermite Kampé de Fériet polynomials for various variables x and y, but, unfortunately, we could not find a regular pattern. However, the following special cases showed regularity. Through numerical experiments, we will make the following series of conjectures.

If $y > 0$, we can see that $H_n(x, y)$ has $Re(x) = 0$ reflection symmetry. Therefore, the following conjecture is possible.

Conjecture 1. *Prove or disprove that $H(x, y), x \in \mathbb{C}$ and $y > 0$, has $Im(x) = 0$ reflection symmetry analytic complex functions. Furthermore, $H_n(x, y)$ has $Re(x) = 0$ reflection symmetry for $y < 0$.*

As a result of investigating more n variables, it is still unknown whether the conjecture is true or false for all variables n (see Figure 1).

Conjecture 2. *Prove or disprove that $H_n(x, y) = 0$ has n distinct solutions.*

Let's use the following notations. $R_{H_n(x,y)}$ denotes the number of real zeros of $H_n(x, y)$ lying on the real plane $Im(x) = 0$ and $C_{H_n(x,y)}$ denotes the number of complex zeros of $H_n(x, y)$. Since n is the degree of the polynomial $H_n(x, y)$, we have $R_{H_n(x,y)} = n - C_{H_n(x,y)}$ (see Tables 1 and 2).

Conjecture 3. *Prove or disprove that*

$$R_{H_n(x,y)} = \begin{cases} n, & \text{if } y < 0, \\ 0, & \text{if } y > 0, \end{cases}$$

$$C_{H_n(x,y)} = \begin{cases} 0, & \text{if } y < 0, \\ n, & \text{if } y > 0. \end{cases}$$

Funding: This work was supported by the National Research Foundation of Korea (NRF) grant funded by the Korea government (MEST) (No. 2017R1A2B4006092).

Acknowledgments: The authors would like to thank the referees for their valuable comments, which improved the original manuscript in its present form.

Conflicts of Interest: The authors declare no conflicts of interest".

References

1. Andrews, L.C. *Special Functions for Engineers and Mathematicians*; Macmillan. Co.: New York, NY, USA, 1985.
2. Appell, P.; Hermitt Kampé de Fériet, J. *Fonctions Hypergéométriques et Hypersphériques: Polynomes d Hermite*; Gauthier-Villars: Paris, France, 1926.
3. Erdelyi, A.; Magnus, W.; Oberhettinger, F.; Tricomi, F.G. *Higher Transcendental Functions*; Krieger: New York, NY, USA, 1981; Volume 3.
4. Gala, S.; Ragusa, M.A.; Sawano, Y.; Tanaka, H. Uniqueness criterion of weak solutions for the dissipative quasi-geostrophic equations in Orlicz-Morrey spaces. *Appl. Anal.* **2014**, *93*, 356–368. [CrossRef]
5. Guariglia, E. On Dieudonns boundedness theorem. *J. Math. Anal. Appl.* **1990**, *145*, 447–454. [CrossRef]
6. Kang, J.Y.; Lee, H.Y.; Jung, N.S. Some relations of the twisted q-Genocchi numbers and polynomials with weight α and weak Weight β. *Abstr. Appl. Anal.* **2012**, *2012*, 860921. [CrossRef]
7. Kim, M.S.; Hu, S. On p-adic Hurwitz-type Euler Zeta functions. *J. Number Theory* **2012**, *132*, 2977–3015. [CrossRef]
8. Kim, T.; Kim, D.S.; Kwon, H.I.; Ryoo, C.S. Differential equations associated with Mahler and Sheffer-Mahler polynomials. To appear in Nonlinear Functional Analysis and Application.
9. Kim, T.; Kim, D.S. Identities involving degenerate Euler numbers and polynomials arising from non-linear differential equations. *J. Nonlinear Sci. Appl.* **2016**, *9*, 2086–2098. [CrossRef]
10. Ozden, H.; Simsek, Y. A new extension of q-Euler numbers and polynomials related to their interpolation functions. *Appl. Math. Lett.* **2008**, *21*, 934–938. [CrossRef]
11. Robert, A.M. A Course in p-adic Analysis. In *Graduate Text in Mathematics*; Springer: Berlin/Heidelberg, Germany, 2000; Volume 198.
12. Roman, S. *The Umbral Calculus, Pure and Applied Mathematics*; Academic Press, Inc.: New York, NY, USA; Harcourt Brace Jovanovich Publishes: San Diego, CA, USA, 1984; Volume 111.
13. Simsek, Y. Complete Sum of Products of (h, q)-Extension of Euler Polynomials and Numbers. *J. Differ. Equ. Appl.* **2010**, *16*, 1331–1348. [CrossRef]
14. Ryoo, C.S. Differential equations associated with generalized Bell polynomials and their zeros. *Open Math.* **2016**, *14*, 807–815. [CrossRef]
15. Ryoo, C.S. Differential equations associated with the generalized Euler polynomials of the second kind. *J. Comput. Appl. Math.* **2018**, *24*, 711–716.
16. Ryoo, C.S.; Agarwal, R.P.; Kang, J.Y. Differential equations associated with Bell-Carlitz polynomials and their zeros. *Neural Parallel Sci. Comput.* **2016**, *24*, 453–462.

MDPI

St. Alban-Anlage 66

4052 Basel

Switzerland

Tel. +41 61 683 77 34

Fax +41 61 302 89 18

www.mdpi.com

Mathematics Editorial Office

E-mail: mathematics@mdpi.com

www.mdpi.com/journal/mathematics